Reclaiming Reason

Reclaiming Reason

*A Christian's Guide
to Recognizing Logical Fallacies*

ADAM MURRELL

RESOURCE *Publications* • Eugene, Oregon

RECLAIMING REASON
A Christian's Guide to Recognizing Logical Fallacies

Copyright © 2012 Adam Murrell. All rights reserved. Except for brief quotations in critical publications or reviews, no part of this book may be reproduced in any manner without prior written permission from the publisher. Write: Permissions, Wipf and Stock Publishers, 199 W. 8th Ave., Suite 3, Eugene, OR 97401.

Resource Publications
An Imprint of Wipf and Stock Publishers
199 W. 8th Ave., Suite 3
Eugene, OR 97401
www.wipfandstock.com

ISBN 13: 978-1-61097-781-4

Manufactured in the U.S.A.

Unless otherwise indicated, all Scripture quotations are from The Holy Bible, English Standard Version® (ESV®), copyright © 2001 by Crossway, a publishing ministry of Good News Publishers. Used by permission. All rights reserved.

Scripture marked "NKJV" taken from the New King James Version. Copyright © 1982 by Thomas Nelson, Inc. Used by permission. All rights reserved.

For my teeming brood,
that you may discern truth from error

Come now, let us reason together, says the Lord . . .

—Isaiah 1:18

Contents

Preface to the Reader ix

1   The Times They Are A-Changin' 1
2   The Language of Emotions 6
3   The Quest to Shape Public Opinion 34
4   What Does That Have To Do with the Price of Tea In China? 51
5   A Straw Man Never Fights Back 70
6   Why Assumptions and Incorrect Inferences Get Us into Trouble 85
7   What's the Future Likelihood? 94
8   If Only It Were That Simple 104
9   *Adieu* To You and You and You 115

Summary of Fallacies and Rubbish 119
Bibliography 123

Preface to the Reader

WHY STUDY logic? As I began to ponder this question, I could not escape the conclusion that one of the most important gifts I can provide to my children is to prepare them to confront the world with a firm grasp of critical thinking skills. If they lack this refined reasoning skill—the ability to think lucidly through all issues—they will be, as the apostle wrote, among those who remain spiritually immature, "tossed to and fro by the waves and carried about by every wind of doctrine, by human cunning, by craftiness in deceitful schemes" (Eph 4:14). Conversely, if my children can logically embrace what they believe and hold firmly to a reasonable faith, they will be well-suited to answer any person who asks them about the basis of their hope; they will be capable of ably defending their Christian faith from a myriad of attacks that are certain to come their way.

Logic is also essential to analyze other people's beliefs. Logical fallacies and poor reasoning flourish in our society. If my children cannot discern fact from fiction—if I have not provided them with the proper and necessary tools to succeed on this most fundamental level—how will they ever be able to differentiate right from wrong, wisdom from folly? Studying logic, even at the rudimentary level, is important to develop and sharpen those God-given reasoning faculties we all possess and need to mature.

Logic is necessary to understand and to convey to others our personal beliefs. Take, for instance, the apostle Paul, who repeatedly went to his fellow kinsmen in every town he visited and expounded the Scripture in a logical and coherent manner. "And Paul went in, as was his custom, and on three Sabbath days he *reasoned* with them from the Scriptures" (Acts 17:2, emphasis mine). The apostle was able to demonstrate to his audiences where they were being inconsistent. Paul would not have been successful in his approach had he not used sound reason.

Finally, logic is necessary to function. Whether or not we are consciously aware of it—or even if we attempt to deny its existence—the truth remains that without the laws of logic and the verbal tools given to us from God, there would be no ability to communicate vibrant and engaging prose: "Tree existential river green buy zogbew tomo*&!^ . . ." What does this even mean? Nothing. It is complete nonsense. The mere stringing together of letters and characters is meaningless apart from logical order.

As human beings who are capable of rational thought, we rely upon following unalterable laws inherent in our constitution and makeup. These laws are binding upon us and ingrained in our very being because of our Maker. We are created in God's image. That is to say, God created us with the ability to reason so he could communicate with us and so we could communicate with others. This logical ability, which is something that stretches into every other facet of life, allows us, most importantly, to take the good news of salvation to those who have ears to hear the truth about God.

For Christian parents, the study of logic cannot be overstated. Understanding the basics and teaching them to our children equips them with lifelong skills they will use daily. For this reason alone, logic should be considered indispensable to every Christian's education, regardless of age. It is never too late to develop rational thinking skill sets. Here, then, is an introduction to *practical logic*, a compendium of common fallacies you are likely to encounter on television, on the radio, on the Internet, in print, or in your workplace.

Newport, Rhode Island
October 2011

1

The Times They Are A-Changin'

BEGINNING IN the 1950s, the Gallup organization began to ask Americans if they believed religion could answer all or most of the problems of the time or if religion was largely old-fashioned and out of date. When Gallup first posed this question to an American audience in 1957, only 7 percent indicated religion was old-fashioned. Since then, the percentage has climbed. It was generally at or around 20 percent during most of the 1970s, 1980s, and 1990s but has spiked to nearly 30 percent in 2008.[1]

The results of the Gallup study are as sobering as they are revealing. Nearly one-third of all Americans—or approximately *one hundred million* people—are persuaded religion is old-fashioned and out of date. Despite the alarming rate at which people are becoming jaded toward Christianity, perhaps the most disquieting figures come from other published studies. These studies show the rapidly dwindling number of people who still claim that religion can answer life's problems seem to have a worldview that is closely analogous to their nonreligious counterparts. For example, other recent Gallup studies revealed Christianity

---

1. Saad, "Americans Believe Religion is Losing Clout," 6.

had little or no impact on the lifestyle choices or values of those in the American culture. Studies show that the difference between professing evangelicals and non-evangelicals concerning abortion, sexual ethics, adultery, and divorce are negligible. In other words, the Gallup polls suggest the Christian religion is making little difference in people's lives; the professing Christian and the non-Christian share a similar worldview.[2] To what extent these studies reflect reality is a matter of debate.

What we can take away from Gallup's study is that many professing Christians—the number of which is unsettled—ostensibly have not given serious thought to the foundation of their beliefs or reasoned through the logical ramifications of their worldviews. Had they spent more time in solemn reflection, the poll numbers would presumably reveal a different story. Without clear and sensible thinking, Christians will scarcely outgrow the current spiritual stagnation and the immature worldview characterized by those who are indifferent or apathetic toward finding consistency between biblical truths and everyday practice. If Christians and non-Christians are like-minded on issues of morality, then we have cause for concern.

So what should characterize a consistent Christian believer? What does that person look like? I submit that a spiritually mature person seeks to know the things of God more intimately and applies biblical principles and understanding to every facet of life. The spiritually developed

---

2. R. C. Sproul writes about his participation as one of several theologians who were asked to examine Gallup's polling data and what it reveals about our society. See the introduction to his *Essential Truths of the Christian Faith*.

Christian should bring *all* thoughts captive to the Word and allow Holy Writ to guide his or her every decision. We start this ongoing process by first knowing what the Bible says and then applying those truths to everyday situations, even the most mundane affairs. In short, mature Christians persistently strive to live a life consistent with biblical teachings by thinking God's thoughts after him through drawing sound, logical conclusions.

This book is your small step in the right direction. It offers pithy examples of logical fallacies and bad arguments to whet your appetite for further, deeper study. This book shows you how to respond to nonbelievers and likewise furnishes you with a new perspective so you will be able to answer Christians when they are being inconsistent regarding their beliefs.

Before we explore any fallacies, we need to point out some general observations. These are not particularly original, nor are they innovative; they are merely descriptive of patterns we are up against when trying to engage in a logical conversation. Let's call these the Ten Commandments of poor reasoning or truisms to keep in mind in our study. For instance, people:

1. tend to be imprisoned to their own traditions (or the one who denies he has traditions is the one most likely enslaved to those very traditions)
2. oversimplify complex issues
3. use emotion when thinking about an issue
4. are not good listeners; they hear selectively and formulate a response without considering your argument dispassionately

5. get side-tracked easily
6. are willing to accept a position that "feels" right without thinking through the potential ramifications
7. judge based on appearances and preconceived notions
8. know less about a topic than they are willing to admit
9. are highly inconsistent in their argumentation and thought process
10. do not always mean what they say[3]

Of course, we could add many more observations to the ones listed above, but these ten declarations are merely given to highlight the areas we will cover throughout this work and to point out the reality against which we all struggle. It is a continual effort because subjective emotions cloud our better judgment. Granted, some are better at controlling their passions than others, but the reality is that we must all think objectively, for the untrained mind generally leads us to faulty conclusions stemming from false premises.

Are people hopelessly entrapped in logical fallacies? Only if they want to be. Clear and rational thinking requires a certain rigor; it, like everything else in life, takes hard work and practice. One must always remain vigilant in conversations, listening closely to another's statements. It requires training, patience, and practice. Yet before one can appreciate the use of reason and examine another's argu-

---

3. For a fuller discussion and listing, see Gula, *Nonsense*, 1–3, and Brown, *Techniques of Persuasion*.

ments properly, one must be aware of the potential pitfalls that await the untrained mind. Therefore, this book reveals the logical traps people employ in the realm of theological or religious discussions. It is a summary of the devices people resort to in order to disguise their untenable positions. Once we understand what we're up against, we will be better equipped to avoid fallacies and assist others in reasoning properly as well.

Time is short, and we have a monumental task before us, so let's begin.

2

The Language of Emotions

Aesop, the ancient storyteller, wrote of an argument that arose between the wind and the sun over which one was stronger. Noticing a traveler below, they decided to settle the matter by seeing which one could cause the man to remove his outer coat. The wind tried first, sending a cold, powerful blast, which nearly knocked over the man and almost stripped the coat of its fastenings. But the stronger the wind blew, the tighter the man seized his coat. The sun decided on a different approach. He dispelled the sky's cloud covering and concentrated his beams of warmth directly on the traveler. The man soon relaxed and unbuttoned his coat. Finally the welcoming rays encouraged the man to remove his covering, and overcome by the sun's radiance, he removed the coat and headed for the nearest shade along the path.

The moral of the fable is fairly obvious: warm persuasion is more effective than cold force.

Aesop's truism has since been echoed by notable playwrights, pundits, and politicians. William Shakespeare's character in *Hamlet*, Rosencrantz, for example, declared, "Many wearing rapiers are afraid of goose-quills and dare scarce come thither."[1] Edward Bulwer-Lytton famously captured the line in his play: "Beneath the rule of men entirely great, the pen is mightier than the sword."[2] And Thomas Jefferson expressed a similar sentiment to his friend Thomas Paine in a letter: "Go on doing with your pen what in other times was done with a sword."[3] These men were merely conveying the idea that words expressed appropriately can have a greater and longer-lasting impact than silencing a dissenting opinion or suppressing a contrary belief. Words stir passions and inflame emotions, creating movements of change, which may either result in accomplishments for the common good or cause interpersonal devastation.

For a positive example, consider the American Revolution. It spawned a republic on the North American continent that enshrined religious liberty in her constitution, enabling generations of Christians to practice their religion without reprisal or fear of persecution. In a negative sense, consider the irrational and unreasonable emotions that overcame sound reason and resulted in the chaos and the nihilistic slaughter of the French Revolution. In one instance, emotions were used for good, and in another context, emotions spawned disaster.

Emotions, then, are a very delicate thing, the manipulation of which can lead either to positive success or to abject failure. Emotions are slight and sensitive, easily tampered with, and amenable, so someone who understands us can manipulate us into doing something or believing something that is untrue by exploiting our emotions.

1. Shakespeare, *Hamlet*, 91.
2. Bulwer-Lytton, *Richelieu*, 89.
3. Jefferson, "To Thomas Paine Philadelphia, June 19, 1792."

The following arguments are emotional appeals designed to prey upon our state of mind, directing our behavior and affecting our motivation, positive or negative. If we are able to recognize such methods, we will be better equipped to protect ourselves from deceit and are less likely to be manipulated into accepting specious arguments.

One of the most commonly used and most effective arguments is the *appeal to pity*. Instead of offering carefully crafted reasons, facts, and evidences for a specific position, a person appeals to one's innate sense of pity and compassion toward others. Some use this tactic in an attempt to reason toward atheism:

> This year's devastating tornadoes are proof the biblical God lacks compassion; he kills and brings pain to the innocent and encourages his followers to do the same. Simply put: the Bible tells us God has the power to protect people but chooses not to do so. Indeed, disease and natural disasters are further evidence that God is either not compassionate or, more likely, he does not exist.[4]

Myriad questions immediately come to mind, especially the numerous unspoken assumptions contained within the statement. A few comments are in order. To begin with, we note how the objector casually dismisses even the mere possibility of the Christian God without offering any objective proof.[5] If this is truly the position the atheist

---

4. See Foreman, "Problems with God: Disease and Disaster" for an atheist's perspective on religion and morality.

5. This is the point where many stumble. Christians should never permit nontheists to get away with assuming the nonexistence of God and demanding Christians prove the contrary. This foundational

wants to embrace, I suggest we focus on what remains after he has completely eradicated God and along with him any objective standard. He has removed the orange tree from the orchard but still wants orange juice every morning. He has plucked all the flowers from the garden but still desires roses on his table every day. I want him to justify his use of reason (the very faculty he uses to argue against his Maker). If there is no God, how can we even have a reasonable discussion about truth or morals?

Let's move on to some other problems with this argument. Making the bold assertion, "God does not exist," is truly an overly simplistic appraisal of the situation, for the reality is far more complex, especially if we were to address some of the underlying suppositions involved in the objection. Consider an important one.

Contained in the objection to God's existence is an assumption of basic human innocence. The theory goes that since most people are genuinely good people, a loving God would never permit harm or disaster to come our way. However, who says we are all intrinsically good and morally upright? Instead of looking at our neighbors and comparing ourselves to them, suppose we regarded our true condition in light of God—a thrice-holy God, to be exact (Isa 6:3). Doing so might help us to see sin for what it is and help us to understand we are not as innocent as originally thought. In fact, having a proper understanding of our nature will aid us in properly seeing our desperate condition apart from divine rescue.

---

atheistic presupposition must always be challenged. Failing to do so places the Christian in a precarious situation and prevents the strongest arguments for God's existence to be discussed.

Next we must ask why a loving God is prohibited from allowing calamity to strike or why he is not supposed to exercise righteous judgment against rebellious sinners, those who, by nature, despise him (Rom 8:7). Instead of assuming only an immoral deity would cause disaster to strike, what if we viewed misfortunes from a radically different perspective? What if God is truly showing his compassion by reminding us, through natural calamities, of a future, certain judgment? Would it be so unreasonable to assert naturally destructive occurrences could be used as a gentle reminder for people everywhere to repent and believe before it's too late? Is this too great a possibility, especially if judgment is to be weighed upon the earth?

Let's also consider another aspect of calamity. It might very well be that God has ordained certain disasters to befall us, on occasion, to bring about a greater good—something that would not be possible apart from evil. For a biblical example of this truth, I remind the reader of the life of Joseph. God ordained Joseph's bondage in Egypt as a slave, and later his imprisonment, so later he would find favor with Pharaoh and would be able to help rule over the land. God ordained the events leading up to and including Joseph's odyssey into Egypt to save scores of thousands of people from famine (Gen 50:20).

Next we have the greatest instance of evil ever committed: the death of Jesus Christ. The crucifixion of Jesus stands as *the* greatest example of God ordaining and controlling evil—albeit through the use of secondary causes—to spare humanity the eternal consequences of a separation from him. Apart from this greatest act of evil, we would have no future hope, so it is very much plausible to point out extraordinary circumstances can result from much unpleasant, seemingly futile suffering.

Finally, the objector ostensibly fails to factor in God's merciful provision and protection for those spared from the ravages of disease or natural disaster. We rarely hear of the success stories of those who narrowly escape the clutches of death or catastrophic events (partly because we just don't know how many times God actually spares us from harm). Occasions on which entire communities or nations are spared are often attributed to blind chance or good fortune.

As popular as the appeal to pity is among atheists, it is not merely limited to attacks against the concept of theism. Evangelicals are just as prone to use this form of argument when advancing certain theological beliefs. Maybe you've heard opponents to the traditional view of hell and the afterlife probe, "How can a loving God send anyone to hell?" We are asked, perhaps somewhat rhetorically, "Does God really consign humanity to unending pain and torment for those who didn't say the right prayer or believe the right thing?" Or worse: "What about someone who has never even heard of Christ?"

An unmistakable feature of all these statements is their effectiveness at producing anxiety. No one wants to think in terms of an unloving or unjust God who sadistically punishes sinners like a crazed madman, but are these thought-provoking questions legitimate? I don't think they are, and here several reasons why.

First, when I hear statements like these—and I'm sure you have heard many similar ones yourself—I am prompted to ask an obvious question: why would any person suppose he or she has the right to define, from a human (finite) perspective, what love is? Confronting this seemingly tough issue of God's character and how he relates to his creation is foundational. Undeniably, one of the key flaws inherent in this line of reasoning is that it fails to account for the true and proper sense of love—that is, the biblical definition. God is not our peer to be judged, and we should never delude ourselves into thinking otherwise.

Another aspect is the misguided assumption that people are banished to hell for *nonbelief*.[6] On the contrary, humanity already stands condemned in Adam[7] (Rom 5:12–14), are born with an innate hatred toward God (Rom 8:7), are dead in trespasses and sins (Eph 2:1), and naturally follow after the devil (Eph 2:2). Against this sobering backdrop, we must understand the only escape from this

6. This charge is not limited to atheists. Some professing evangelicals make similar assertions. For a contemporary evangelical who makes this argument, see Bell, *Love Wins*.

7. Federal headship is the biblical teaching that all humanity was represented by our first parents in the Garden of Eden. Specifically, Adam acted as a representative for the entire human race, so when God set before Adam and Eve a trial, he was testing the entire human race. Adam—whose name means "man" or "mankind"—was chosen by a perfectly holy, perfectly just, and omniscient God to act as the head of all humanity.

The consequence of Adam's federal headship means that when Adam sinned, he sinned for the entire human race. His transgression was our transgression; his punishment was our punishment as well. It was not just Adam and Eve who were cursed with the sweat of the brow and labor pains in childbirth. The same is true of us all today. Bear in mind that no other person in human history was a better, more perfect representative than Adam, and yet he still fell, bringing sin and death to the world. This is precisely Paul's meaning when he says, "that through the disobedience of one man, death comes into the world." Being "in Adam" brings death, but when they are represented by the second Adam (Christ), believers are made alive.

poignant reality is to repent of sin and believe in Christ Jesus. Contrary to popular opinion, no person is consigned to everlasting perdition for nonbelief. Rather, those who find themselves sentenced to everlasting destruction will understand their just condemnation is the consequence of guilt incurred from personal sin.

The final question raised concerning people who never heard the gospel call to repentance assumes God is under obligation to offer all humanity an equal opportunity for salvation. The reality, though, is that the Lord is not obliged to his creation in this manner. I don't mean to sound flippant, and I understand this might sound harsh to some, but I would encourage us to think beyond the limited way we normally process information and consider the bigger picture, if you will. The bottom line is that God is not indebted to his creation; he does not owe any single person a chance to be saved. That's why salvation is referred to as *free grace*. For those who are less easily consoled, we must remember that we don't even know exactly how God will handle issues such as those who have never heard of the Bible or the gospel. If young children and the mentally deficient are not accountable before God, it is not our appointed duty to decide for God exactly which humans are at a point of accountability. We cannot attempt to take God's place and decide these issues for him.

Sometimes the nontheist objects that a "sincere searcher for God" or an "innocent young teen" would perish forever simply because he or she has no awareness of the truth. "Only a cruel and vindictive god," he protests, "would punish someone with seventeen billion years in hell for only seventeen years of sinful living. Such teaching is clear proof the Christian God does not exist!"

Okay, so let's tackle this common objection. What is the single greatest unspoken assumption in this statement? If you said that people only commit sins in this life and cease doing evil in the next, you would be correct. Indeed, why should we believe that sinning ends once people die (since, after all, death does not equal nonexistence)? Think about this. If the restraining hand of God is removed from the denizens of hell, then it is reasonable to assert that their hatred and obstinacy toward their Maker will only be magnified; they will in all likelihood be more established in their wickedness in the afterlife since there is no longer any divine restriction. When God removes his protection and restraining power, their absolute depravity will be made abundantly manifest because they will be incapable of doing anything other than sinning against God.[8] On this basis, it is dubious to argue people are punished in perpetuity for a (relatively) few years of existence.[9]

---

8. John Gerstner makes the following argument about the inhabitants of hell: "That is, these unchanged sinners who hate God will hate and curse God for the punishing. That certainly increases their sin and incurs more punishment. Their guilt, in other words, is constantly increased and so is their deserved punishment" (Gerstner, *Repent or Perish*, 74).

9. Of course, this leads us to another controversial question: Does God purposefully bestow immortality (an attribute that belongs to God alone, 1 Tim 6:16) to the inhabitants of hell for the sole purpose of punishing them forever, or does the Bible teach that the wicked will finally, truly, and ultimately perish and become extinct forever after God imposes some degree and duration of conscious punishment? That truly is a question worth exploring.

A variation of the appeal to pity, which is perhaps sometimes indistinguishable from it, is the *plea for special treatment*. This form of argumentation is especially prevalent among liberal evangelicals and folks who oppose the exclusivistic claims of Christianity. One of the boldest expressions of the plea for special treatment is found in Rob Bell's book *Love Wins*. The emergent church pastor relates the story of a woman who, after viewing a painting with a quote from Mahatma Gandhi, scribbled the following words on a piece of paper: "Reality Check: He's in hell."[10] Bell was aghast.

"Gandhi's in hell?" he wondered rhetorically. "We have confirmation of this? Somebody knows this? Without a doubt?"[11]

Bell has a point. He is correct in one sense; we don't have confirmation of Gandhi's ultimate fate. And quite frankly, Christians should never make it a practice to make dogmatic statements about people's eternal destinies, but what we are aware of, and what we can speak confidently to, are Christ's words regarding the pathway of redemption. Jesus proclaimed that he is the *only* way to the Father (John 14:6). Yet Bell's words lead us to the assumption he believes God granted Gandhi a special dispensation from the biblical requirement of repentance and belief in Christ as personal Savior to obtain everlasting life.

But we ask: Should God, in opposition to his own revelation in Scripture and sense of divine justice toward violators of his law, grant special favor toward, in this instance, Gandhi for virtuous deeds accomplished? What are we to

10. Bell, *Love Wins*, 1–2.
11. Ibid.

do with the prophet Isaiah's words when he declared even our seemingly righteous works are no better than "filthy rags" (Isa 64:6)? Again we return to our previous discussion about God's holiness and our sinfulness in light of such majesty.

From this perspective, is Gandhi—or are any of us—really all that deserving of heaven? Can we buy the riches of Christ with our "filthy rags"? The point to remember is that assuming people enter into God's presence on the basis of charitable works completed is not a valid reason for asserting that God *must* grant any person eternal life. Otherwise, this puts God in the position whereby he is obliged to his creation, rewarding what is owed to them on the basis of works accomplished instead of imparting free grace to undeserving sinners.

Closely connected with the appeal to pity is the *appeal to guilt*. Let's continue with our example of the Christian doctrine of hell:

> Of the many billions of people who have ever lived, will only a select number "make it to a better place" and every single other person suffer in torment and punishment forever? Is this acceptable to God? Has God created millions of people over tens of thousands of years who are going to spend an eternity in anguish? Can God do this, or even allow this, and still claim to be a loving God?[12]

We are invited to think of the never-ending torment people will experience under God's wrath. This image is

---

12. Ibid. Also, for an exegetical rebuttal to *Love Wins*, see Chan, *Erasing Hell*.

designed to haunt us until we acknowledge the traditional view of hell is merely an archaic, medieval concept or until we abandon the exclusive claims of Christ being the only way to "the better place."

There are several points to contemplate. First, we must point out that no one has the right to prey upon the emotional instability of any other individual; such is the epitome of manipulation. Second, unless the protagonist can offer compelling reasons why the traditional view of punishment and the afterlife should make any Christian believer feel guilty—especially in view of fact that Christianity teaches everlasting life for any person who believes—it remains merely speculation.

Finally, even if we were to feel a sense of guilt for believing in the traditional view of hell, there has been no evidence proffered suggesting that a repudiation of hell produces any efficacious results for nonbelievers. Even if all of humanity rejected the biblical doctrine, that alone would not prove the orthodox understanding of hell untrue. Christians should take comfort in knowing the Bible presents a God who is far more perfect in love and fairness than we humans can even contemplate. "Let us fall into the hand of the LORD, for his mercy is great" (2 Sam 24:14). The Lord Jesus Christ will never do anything that is unmerciful or unjust.[13]

---

13. What exactly is *fair* and *just* has produced division within evangelicalism, leading some as of late to abandon traditionalism in favor of universalism (or some variation thereof). For an alternative view to both positions, known as "conditional immortality," see Fudge, *The Fire That Consumes*.

Another illustration of the appeal to guilt is the argument commonly proffered for legalizing homosexual marriage (by the way, the expression is an oxymoron since marriage is defined by the union of one man and one woman). We are told with much alacrity that homosexual marriage is an intrinsic right that has been bestowed upon us by our Creator, among the same inherent entitlements that guarantee life, liberty, and the pursuit of happiness. People who oppose this "right" are branded as intolerant and are viewed as perpetuating worn-out religious arguments that only serve to deny American citizens equal protection under the law. We are made to feel guilty if we believe God instituted marriage and as a result, strive to uphold a union between one man and one woman. In short, opposing the redefinition of traditional marriage is looked upon as bigoted, selfish, and tantamount to the oppression of African-Americans. This perspective was crystallized by former United States Solicitor General Theodore Olson in an article for *Newsweek* magazine. Olson wrote:

> Legalizing same-sex marriage would also be a recognition of basic American principles, and would represent the culmination of our nation's commitment to equal rights. It is, some have said, the last major civil-rights milestone yet to be surpassed in our two-century struggle to attain the goals we set for this nation at its formation . . .
>
> Over the years, the United States Supreme Court has repeatedly held that marriage is one of the most fundamental rights we have as Americans under our Constitution. It is an expression of our innate desire to create social partnership, to live and share life's joys and

> burdens with the person we love, and to form a lasting bond and a social identity. The Supreme Court has said that marriage is a part of the Constitution's protections of liberty, privacy, freedom of association, and spiritual identification. In short, the right to marry helps us to define ourselves and our place in a community. Without it, there can be no true equality under the law . . .
>
> Americans who believe in the words of the Declaration of Independence, in Lincoln's Gettysburg Address, in the 14th Amendment, and in the Constitution's guarantees of equal protection and equal dignity before the law cannot sit by while this wrong continues. This is not a conservative or liberal issue; it is an American one, and it is time that we, as Americans, embraced it.[14]

His urging might sound persuasive to some, but let's think beyond the emotional pleadings and empty rhetoric for just a moment. For starters, why should marriage be restricted to the union of two people and not between three, four, five consenting adults or more? Upon what basis do we define marriage as between two people—and only two? Why can't I marry my sister (brother, mother, father, etc.)? Why is marriage limited to human beings? If my pet is all I have and I'm in love with my pet, why not? Or in the wake of scores of centuries of human civilization, why now do we think we are more enlightened than previous generations and should overthrow the traditional understanding of marriage?

14. Olson, "The Conservative Case for Gay Marriage," 28ff.

These are just a few of the many questions pro-homosexual advocates should be pressed about in order to have a robust and meaningful dialogue. Additionally, there are other uncomfortable realities that many have probably never considered, such as the shared benefit to society only traditional married couples can offer—namely, the ability to provide offspring for the society's continuation. Instead of interacting with these fair concerns, we are insulted and made to feel guilty if we respect the boundaries established by God by rejecting all forms of perversion to the divine institution of marriage.[15]

Another form of argument used is the *appeal to fear*. The appeal to fear is an attempt to alarm someone into accepting a specific belief. "If you do X, then Y will inevitably happen," the argument goes. (Or conversely, "If you don't do X, then Y will inevitably happen.") Of course, Y is something dreadful that no one would want to come to pass, such as, "If abortion is outlawed, women will be forced to seek unsafe, back-alley abortion doctors." However, the speaker must first demonstrate a logical connection between X and Y; one must establish a proper cause-and-effect relationship between the stated premise and the assumed conclusion. In this instance, we were never told why women must seek out abortions—as if there is no other alternative—just that they will choose murder instead of protecting and respecting life.

To offer another example, the sixteenth-century Dominican monk John Tetzel was notorious in Germany for selling papal indulgences—that is, remission of sins to

---

15. For a chronicle of the dramatic cultural changes that have taken place in the United States in relation to homosexuality, see Brown, *A Queer Thing Happened to America*.

escape the flames and torment of purgatory. Tetzel and his preachers promoted their wares by using notable catchphrases such as, "Indulgences make sinners cleaner than when coming out of baptism" and "Cleaner than Adam before the Fall." Citizens contemplating the purchase of indulgences for deceased loved ones were promised, "As soon as the coin in the coffer rings, the soul from purgatory springs."[16]

Their message resonated: give money to the Catholic church or risk the flames of purgatory for you and your family. Had some of Tetzel's unsuspecting victims been more shrewd, they would have realized his claims were not sound, since he had no authority or basis for declaring that monetary offerings release souls from the temporal punishment due to sin (not to mention his requirement to establish from Scripture the legitimacy of purgatory).

Sometimes the appeal to fear is personally directed. Political maneuverings of the papacy during past centuries offer abundant examples. One such case is the promulgation of the papal bull *Unum sanctum* issued by Pope Boniface VIII in 1302 after he quarreled for some time with the king of France. Boniface, in that document, declared in part: "There is one holy Catholic and Apostolic Church, outside of which there is neither salvation nor remission of sins," and "it is altogether necessary to salvation for every human creature to be subject to the Roman Pontiff."[17] The French sovereign received the not-so-subtle message that he should submit to the authority of the pope or face the eternal consequences. Nothing could be more dreadful than spending an eternity in

16. Gonzalez, *The Story of Christianity*, 21.
17. Bettenson, *Documents of the Christian Church*, 115–16.

flames, so the pope's threats of pain and suffering toward the impenitent proved successful. The king of France ultimately complied with the pope's demands.

A more recent example—something we are witnessing with greater regularity all the time—involved Dale McAlpine, a Christian street preacher in England. McAlpine was arrested last year in the Cumbrian town of Workington and charged with a public-order offense after telling a passerby homosexuality is a sin. After the woman walked away and reported the exchange, a policeman approached the evangelist and identified himself as the "Lesbian, gay, bisexual, and transgender liaison officer." McAlpine responded, "It's still a sin."[18] The evangelist was subsequently detained and prosecuted by the sexual revolutionaries, being charged with using "abusive and insulting words," contrary to the government's existing laws.

In this case, the charges were eventually dropped, but the Baptist minister and other would-be evangelizers received a vibrant signal, a tacit message that was clearly intended to curb overt preaching against deviant sexual practices. What we also discovered from this episode is the new repression that claims "mutual tolerance"—but doesn't resemble anything like it—is spreading rapidly. Indeed, this process of appealing to fear, if not checked, will lead in the end to the arrival of the Thought Police at a community near you.

Nonetheless, if scare tactics prove unsuccessful, praise might just achieve the desired result by employing the *appeal to flattery*. Let's face it—we all enjoy words of encouragement and adulation. When we are flattered, though, we tend to confuse our optimistic feelings about the person

18. Anonymous, "Christian preacher on hooligan charge," 71.

with what the person is actually saying. For example, Beth and John are engaged in a discussion, and Beth makes excessively flowery statements about John's work in the field of theology. John is pleased with Beth's words; therefore, John holds a favorable opinion of Beth, making him more agreeable toward Beth's statements. Yet Beth hasn't offered any valid reasons for why her position is true and why John should embrace her opinion.

The *appeal to hope* takes this form: "If you do X, Y might happen; therefore, if you want Y to happen, do X." It must be pointed out that doing X does not guarantee Y will happen, nor is compelling evidence offered that leads one to believe it will happen. "It's all a matter of faith. You have one-dollar faith, and you ask for a ten thousand-dollar item; it isn't going to work. It won't work. Jesus said, 'According to your faith,' not 'according to his will, if he can work it into his busy schedule.' The Lord made the promise for us all, 'according to your faith be it unto you.' Now I may want a Maserati but have only a bicycle faith. Guess what I'm gonna get? A bicycle." Even if you have "Maserati faith," you may not receive a Maserati because God promises to supply our needs, not simply our dreams or wants.

Another renowned televangelist wrote a book about divine healing that contained the following:

> I am fully convinced—I would die saying it is so—that it is the plan of our Father God, in His great love and in His great mercy, that no believer should ever be sick; that every believer should live his full life span down here on this earth; and that every believer should finally just fall asleep in Jesus.[19]

19. Hagin, *Seven Things You Should Know About Divine Healing*, 21.

Another promised:

> God wants to do something supernatural in your life; he wants to do something supernatural in your finances. Can you believe it? God is interested in all your needs, wants, desires, and even, your prosperity. In fact, the psalmist expressed that God delights in the prosperity of his people (Ps. 35:27). All you must do is receive God's Word into your spirit by agreeing with him, and then you can expect God to move in your finances! I encourage you to pray this prayer: "God, I receive Your Word, and I agree with You. I expect supernatural blessings and favor for my family. I expect it, I look for it, I declare it in faith, and I receive it in faith, in Jesus Name, Amen."[20]

Often our sense of hope—hope that yearns for financial security and splendid health—overrides sound judgment, making us forget just how improbable—and biblically untrue—certain hope claims are.[21]

The *appeal to sincerity* is quite effective and very persuasive if presented by the right person under the right circumstances. Consider the following events. Sarah sits silently on the stairs, listening intently to a Christian evangelist attempt to explain the evolution of Mormon theology and try to reason with her father concerning key tenets of

---

20. Comes, "Can You Perceive It?," 27–29.

21. For a helpful study outlining and articulating the abuse of the Bible from those within evangelicalism, see Brauch, *Abusing Scripture*. Professor Manfred Brauch argues mistaken interpretation and application of Scripture is a detriment to the integrity of the Christian witness and contributes to profound misunderstandings in Christian belief and practice.

the Mormon faith that, when examined dispassionately, are found to be incongruous with the Bible. Unable to sit silent any longer after witnessing her Mormon father struggle to defend his own beliefs and powerless to muster any compelling response to the overwhelming evidence presented, she confronts the man with an earnest, sincere, and unassuming tone. She initially grimaces and appears to be struggling for the right words to express her heartfelt convictions. She begins to offer her defense, pausing at times to ensure the proper words are expressed. Her sentiment is so profound that words are, at moments, elusive; she repeats certain phrases, though not exactly the same way, for emphasis; and particular emphatic words are used to demonstrate her sincerity.

"Beginning with my great-great-grandparents who accompanied Joseph Smith on his voyage across America in search of the New Jerusalem, all of my family members have been Mormons, so naturally, I was raised as a Mormon as well. But then there came a time in my life—as there probably is in every person's life—when I started to question the doctrines my parents taught me. During my early teenage years, I enjoyed attending church and participating in all the activities that accompanied the church, but I wasn't fully convinced if everything I had been taught was true.

"Specifically, I wanted to know, either way, whether the Book of Mormon was truly from God or if it was a pious fraud concocted in the fertile imagination of Joseph Smith. I knew it had to be one or the other. I decided I would read the book in its entirety and pray about it. After all, I had learned early on the Book of Mormon promises in Moroni

10:3–5[22] that if I read with real intent and prayed earnestly about the book, I could discover truth through the power of the Spirit. I wasn't exactly sure what to expect. Yet, as I read the book, I started to feel a burning sensation in my heart; a flame was kindled that could not be quenched. I suddenly felt peaceful and happy, and it all started to make sense. Later I realized that the Spirit was testifying to me that what I was reading was divine revelation, that what I was reading was true.

"So, no, Mr. White, my father and I may not be able to fully address your objections or respond to your questions clearly enough to your satisfaction, but I know with absolute certainty that what I believe is true. There is nothing you could say or do to make me change my mind because I know what I experienced, and I know that it was from God. I know because I sensed a burning in my bosom that I never felt before or since about anything else. I believe—I firmly believe that my experience and testimony to the truth of the Book of Mormon is true. I do believe this wholeheartedly. After having examined the evidence carefully, I am convinced your theological position is untenable, and I do wish you would reconsider, that you would pray earnestly

---

22. Moroni 10:3–5 reads, "Behold, I would exhort you that when ye shall read these things, if it be wisdom in God that ye should read them, that ye would remember how merciful the Lord hath been unto the children of men, from the creation of Adam even down until the time that ye shall receive these things, and ponder it in your hearts. And when ye shall receive these things, I would exhort you that ye would ask God, the Eternal Father, in the name of Christ, if these things are not true; and if ye shall ask with a sincere heart, with real intent, having faith in Christ, he will manifest the truth of it unto you, by the power of the Holy Ghost. And by the power of the Holy Ghost ye may know the truth of all things."

like I have, and that you too would be enlightened to the truth of the prophet, Joseph Smith."

The young woman urged Mr. White to abandon his Christian convictions and trust her because she experienced a subjective feeling of certainty, an earnest sensation about the Mormon religion. While there was little doubt in her mind as to the truthfulness of her deep-seated convictions, she never offered logical reasons why Mr. White should accept her testimony as true. After all, it could rightly be pointed out that every person who believes in something could give a personal testimony about the certainty of inner feelings. The Bible, conversely, warns repeatedly against following the desires of an unregenerate heart and permitting it to function as a moral compass. It is described in the most sobering terms: the heart is "deceitful above all things" and "desperately wicked" (Jer 17:9), and, whoever "trusts in his own heart is a fool" (Prov 28:26). God must first grant a new heart before it can properly assent to truth and righteousness and before it can be trusted to serve as a guide and moral compass.

Another emotional tactic can be classified under the generic term the *appeal to friendship* or more specifically the *appeal to love* and its close relative, the *appeal to trust*. Someone tells you that since you don't accept certain personal choices, you don't love her or you're not truly her friend: "If you really loved me, you would accept my decision." This, of course, is an unfair conclusion. Friends or loved ones do not always have to agree all the time. Agreements or disagreements have nothing to do with personal feelings toward an individual.

We would have doubts about a person's discernment if she always accepted a decision, regardless of its magnitude, based solely on the desire of someone else. Not fully understanding the potential consequences of particular actions would make consenting or accepting some decisions reckless and irresponsible. For instance, consider if Monica informs her friend, Alyssa, that she is sexually promiscuous with her longtime boyfriend and seeks her approbation and consent. "You're my friend, and it's your responsibility to accept my decision like a friend and be happy for me." Sometimes relationships necessitate a voice of reason to help prevent unwise decisions that can have enduring negative consequences.

Similar to the preceding appeals is the *appeal to pride* or *loyalty*. For example, someone might declare, "If you truly care about the propagation of the Christian message, if you really want to see Christianity expand, you would stop debating other Christians over nonessential issues." Or he might say, "What do you mean you won't read from the King James Version of the Bible? Are you against the teaching of Christ that he will preserve his Word without error?" The appeal to pride or loyalty, of course, is a glaring oversimplification of an issue. Engaging in intramural debates with fellow believers does not automatically mean you're *not* committed to the expansion of the gospel. Neither does it follow that reading from various English translations renders someone less spiritual, as if the doctrines of inspiration and inerrancy are solely incumbent upon a seventeenth-century translation committee.

The *appeal to the bandwagon* takes advantage of our sense of belonging—the desire to be like our neighbors or to "keep up with the Joneses." We are urged to forsake organized religion since it inevitably leads to division, inequality, and holy wars. Someone might spout off, "A growing number of intellectuals now view modern science as an adequate substitute for God." We are told, "No biblical scholar of any notable repute believes in the divine inspiration of the Christian Bible, let alone its supposed inerrancy." On a reduced, though no less important, level, some claim, "No serious scientist believes in the Genesis account of creation. All academicians and rational-thinking individuals embrace the neo-Darwinian evolutionary theory." With these examples, as is the case with every other emotional appeal, no valid reasons have been given for why one should forsake religion, dismiss the divine nature and integrity of Scripture, or embrace a neo-Darwinian evolutionary theory as scientific fact. The only reason provided is that everyone, at least the so-called respectable individuals, believe accordingly, so we must follow suit.

Another emotional argument often used is the *appeal to status*. Most of us, though some more than others, are sensitive regarding status, whether we like to admit it or not. We cannot escape this fact. How many times have you been around people who "dropped names" of important or influential figures with whom they have made acquaintance? They showcase their new foreign luxury sports car, inadvertently mention they have a maid or housekeeper, or try to dazzle others with their extensive travel itinerary. Whenever they buy something the name on the label is of prime importance. These people are most susceptible to appeals to status, of assuming special claims of import are more true or important.

University of North Carolina professor and *New York Times* bestselling author, Bart Ehrman, seems especially fond of this approach. During a debate regarding the reliability of the New Testament text, Ehrman pressed his opponent, James White, asking him to name some of the current leading textual critical scholars of Europe, individuals with whom Ehrman (as he claims) has much familiarity. Let's look at a snippet of their exchange during the cross-examination portion of the debate:

> EHRMAN: Okay. Would you agree that Eldon Epp is probably the dean of textual criticism in America today?
>
> WHITE: I think Eldon Epp, yourself, and D. C. Parker are probably the biggest names right now, unfortunately I would say that the perspective that you are now pursuing—and as you yourself have said the past 10–15 years—you've pretty much given up on working on the original texts, that's sort of been done—
>
> EHRMAN: So, okay, so Epp in America, and Parker—he's English—and maybe Keith Elliot in England, he's a big name. How about in Germany, who would be the authorities now, living?
>
> WHITE: With the Alands out of the picture—
>
> EHRMAN: Barbara's still living.
>
> WHITE: I'm sorry?
>
> EHRMAN: Barbara's still living.
>
> WHITE: Yeah, but I don't think she's publishing or anything. She's retired from the institute.

EHRMAN: Maybe Klaus Wachtel or Gerd Mink . . .

WHITE: Yes, well, I'm sorry I don't keep up with German textual criticism today.

EHRMAN: How about in France?

WHITE: I don't know anybody in France, sir.

EHRMAN: Probably Christian-Bernard Amphoux.

WHITE: Mm-hmm.

EHRMAN: These are the biggest name in the field: Epp, Parker, Elliot, Aland, Wachtel, Mink, Amphoux. So far as I know, none of them agree with you on this particular point about the preservation of the text.[23]

Ehrman's repeated appeal to his relationships and knowledge of leading textual critics in America and abroad was presented as "evidence" to demonstrate his opponent's position was inaccurate. While having notable connections might impress some, what defines truth is not one's associates or even the number of esteemed people who embrace any given position. Erhman's authority language, informing the audience that he associates with the elite scholars in the field of textual criticism, is designed to give the impression that he is more influential and important than he might actually be. If he can convince the audience he *is* the authority or that he knows the authority in New Testament textual criticism issues, his position, by default, is deemed true.

Elsewhere, for example, someone might say, "If Jim were truly a theological scholar, then he would have a

---

23. For a complete transcript of the debate, see White and Ehrman, "Can the New Testament Be Inspired in Light of Textual Variation?"

doctoral degree from an accredited university. As it is, his studies were completed outside of a brick-and-mortar school. There was no set syllabus, the school has no library, and there are no curriculum committees and no course review procedures." Needless to say, a person should gain credibility by the quality of the work he produces and not by the product (or school, in this instance) he uses. It is just as fallacious to argue that someone graduating from a non-accredited university cannot produce scholarship as it would be to conclude that a graduate from a prestigious institute of higher learning necessarily produces top-level scholarship and truthful findings. Mere appeals to status, perceived status, or the lack thereof are not logical reasons for accepting or rejecting someone's truth claims.

Sometimes people use the *appeal to tradition* as a means of justifying their beliefs. "This has been the belief of the church since the time of the church fathers; therefore, why change now?" Such a statement in and of itself should carry little weight, as multiple stripes from inside and outside of Christendom make similar claims. Most notably, Roman Catholics have for centuries idolized precedent and have argued on the basis of tradition for such doctrines and practices as the papacy, papal infallibility, purgatory, the Mass, the Marian dogmas, and so on. Greater reasons need to be given other than asserting something has a rich pedigree (whether it can be demonstrated throughout history via empirical evidence is another issue altogether). Even if a belief, doctrine, or position can be traced throughout the ages, circumstances change with time, and greater clarity can bring renewed understanding. Sometimes certain beliefs, practices, or customs need to be reformed.

These are just a few of the more predominant examples that make up the broad category of emotional appeals, arguments that are directed toward influencing our feelings. In truth, while sometimes such attempts are intended to manipulate or deceive, it must be pointed out that this is not always the case; there is nothing intrinsically wrong with emotional appeals, especially when they reflect genuine, profound convictions about a particular belief. In the example stated earlier, young woman may truly believe Joseph Smith is a prophet of God and that denying him will result in eternal consequences, and she may regard the appeal to sincerity as a shortcut to get you to do what is necessary for salvation. The emergent church pastor may genuinely believe the vast majority of Christendom has fundamentally misunderstood the nature of hell, and it may manifest itself in the form of an emotional appeal to guilt. The televangelist might truly believe that if you receive God's Word and donate to his ministry, God will, in turn, bless you financially.

What is imperative to recognize is that an emotional appeal may also reflect some underlying issue or ulterior motive that is not openly mentioned or an agenda that is being concealed. Perhaps the person trying to coax you out of debating other Christians is genuinely interested in spreading the gospel message; on the other hand, maybe he is merely trying to avoid having to defend his theological convictions in an open and honest format, knowing his position would dissolve under scrutiny. Look for the hidden agenda, and do not be so quick to judge based on feelings alone because, as we have seen, emotions can be manipulated through seemingly persuasive arguments until critically examined under the soberness of sound reason.

3

The Quest to Shape Public Opinion

WHEN WE hear the term "propaganda," we are tempted to regard the word as evil or sinister. Perhaps images of Nazi Germany come to mind. Or maybe we picture something a bit less ominous, such as a political speech, unruly protestors, a television show, or an advertisement.

All of these are certainly forms of propaganda. In the widest sense, propaganda is any strategy for spreading ideas or beliefs that appeals to our emotions and not to our reason. Propaganda attempts to get us to act or to think in a particular way; it seeks to alter our habits by influencing our beliefs and attitudes through subtle or even deceptive means. Rarely, if ever, is a propagandist open about his or her intentions. Instead, he or she often has a concealed agenda that is cloaked in specious, emotional language. The best propagandists are those who are capable of exploiting basic emotional needs through suggestibility and couched language, winning one's confidence, tampering with emotions, and getting someone to accept as true that which is patently untrue without providing compelling evidence for the conclusion. Propagandists are skilled in manipulating others and controlling their emotions in a way that is designed to make some agree with the propagandist without reasoning sensibly through the logical consequences. Propaganda can be, on many levels, quite dangerous.

Not all propaganda is bad, however. There is nothing intrinsically wrong with spreading ideas or beliefs and encouraging others to embrace our convictions—so long as it is done honestly, nonforcibly, and with pure motives. Sometimes the reasons people employ for us to purchase their product, vote for their candidate, or accept their worldview or theological beliefs are based on emotion and not on clear, rational thinking. A commercial will pressure us to buy a certain truck because it looks tougher, not because it necessarily is tougher; a movie will agitate us to feel troubled about capital punishment without thinking about how the Bible addresses the issue. A pro-abortion advocate will wax eloquent about the fundamental right of a woman's reproductive health without addressing the moral concerns raised by the issue. All of these examples propagandize in some form.

Sometimes emotions correctly lead us into truth, but this is not always the case. Emotions can override sensibility, which is why the New Testament warns against superficial decisions: "Do not judge by appearances, but judge with right judgment" (John 7:24). The biblical idea is uncomplicated. We must delve below the layer of triviality rather than relying merely on our initial emotions and outward appearances—because emotions, after all, can easily be provoked and misled. So when people prey on the emotions of others, it is often difficult to decide whether to follow one's emotional instincts or not. In this chapter, we will discuss how to recognize examples of *manipulative* propaganda and why it is does not constitute good reason for accepting a position.

One kind of manipulative propaganda utilizes *repetition*. The perception is that if someone keeps saying something often enough, long enough, loud enough, and without correction, people will eventually believe it. Propagandists repeat their platitudes, and sooner or later, they become "true." Copious examples of this truism abound in topics such as abortion, evolution, homosexuality, and biblical criticism, to name just a few: "The ability of a woman to have complete control of her own body is critical to civil rights." "When will people stop the silly pretenses and accept the scientific fact that biological evolution has produced the diversity of living things over the billions of years of Earth's history?" "Separation of church and state bans government at all levels from allowing religious expression within public facilities or by official bodies." The latter, of course, is an egregious misreading of the Establishment Clause, and justification for such action is nowhere to be found in or supported by the Constitution of the United States. Yet, with enough time and persistence, all of these contentions have, in many minds, achieved truth status.

Another property of propaganda is *confidence*. As one might expect from propagandists, they speak with great aplomb and an inordinate sense of knowing a topic intimately. Their posture is erect; their voice is commanding; their facial expressions are striking; and their gestures are decisive. They appear to be in complete command of the situation. Therefore, they must be experts on their subject—right? The theory postulates that if individuals are confident in their position and certain that what they are saying is true, they *must* be right.

For example, an esteemed philosophy professor, an intellectual in his own right, discusses, in a chapter in his *New York Times* bestselling book, the process of human development over millions of years under divine guidance. He waxes eloquent about the evolutionary process and how God used it as the mechanism by which the universe exists today. Shawn is impressed with much of the professor's work in other areas, but more importantly, he reasons the lecturer *must* be correct—just look at his poise and the way he presents vast quantities of information and with such rapidity. So Shawn assents to the conclusion offered, embracing it as truth.

What has Shawn actually done? He has accepted as true a position based upon the style of presentation and the mellifluous sound of an instructor discussing matters admittedly beyond his area of expertise. Shawn has also only heard one side of the matter. The ancient adage comes to mind: "The one who states his case first seems right, until the other comes and examines him" (Prov 18:17). Confidence might be a sign one thoroughly understands the position, yet it does not automatically safeguard that the particular position is true.

Another tool of propaganda is *oversimplification*. The propagandist takes one side of an issue while ignoring the counterarguments and treats it as if it were the only perspective. Complex and multifaceted issues are reduced to one of two extremes, occasionally resulting in an *either-or* dilemma. "Either you go along with me and my belief that the King James Bible is the preserved Word of God for

English-speaking people, a conviction that stands for the good of Christendom, or you don't. And that, quite obviously, means you choose the camp of heretics and schismatics who seek to justify using modern versions that pervert the Authorized Version."

It's quite simple, he insinuates. He says all believers should use God's divinely inspired English text, all the while ignoring the fact that modern translations utilize manuscripts (older, better-attested texts) that have been discovered since the original production of the seventeenth-century King James Version. Also missing in his comments were an acknowledgment of objections to "KJV onlyism," including a recognition of the difficulty in translating from one language to another, textual problems in the *Textus Receptus*—the basic Greek text underlying the 1611 Authorized Version—oversights of King James Bibles, differing editions of the same, and the attitudes of those who support it. Likewise, when discussing reasons why people do not embrace KJV onlyism, he may oversimplify the problem: "It's all because of . . . " Needless to say, this is a worn-out tactic that distorts the truth of the matter.[1]

Name-calling or *labels* is another tactic used by propagandists. This represents an instinctive, near-Pavlovian response to an argument individuals don't like, or perhaps a way some reply when they've simply run out of ideas. Propagandists assign abusive epithets or use pejorative

---

1. For a discussion regarding the divine inspiration of the King James Bible, see Carson, *The King James Version Debate*. Also, for an understanding of the historical development of Bible translations and some basic issues of textual criticism, see White, *The King James Only Controversy*.

names that have strong, negative emotional connotations. They expect to influence opinions and attitudes negatively toward their challengers; they hope to shut down discourse with the opposing side before it even begins. Sometimes these terms will include: homophobe, bigot, anti-choice, anti-Catholic, Jesus freak, religious right-wing fundamentalist, and so on. Conversely, they use positive language to describe allies or positions compatible with their beliefs: reasonable, open-minded, fair and honest, sensible, and the like. While terms, flattering or not, might play a role in biasing some in the audience, no one should accept downgrading characterizations until these unkind people first define their terms and provide meaningful evidence to support the stated claims.

Closely related to name-calling or labels is the technique of *poisoning the well*. When an adversary poisons a well, he or she ruins the entire water supply, no matter how pure or good the water was prior to introducing contaminants. The enemy, by his or her deliberate actions, makes the water unusable. When adherents of a certain position introduce toxic information about a person or subject, whether true or not, they are trying to ensure their opponent will either be merrily disregarded or viewed with skepticism. For example, before Mr. Palmer has been introduced to an audience, someone interjects, "Mr. Palmer is a very good speaker, but talking is all he ever seems willing to do. He's always correcting others; he always has to express his opinion." How can Mr. Palmer respond to the charge? If he remains silent, he runs the risk of appearing to accept the criticism. If, however, he defends himself, he's correcting the objector, which would seem to confirm the accusation.

In this instance, the well has been poisoned, placing Mr. Palmer in a precarious situation.

Poisoning the well occasionally applies to concepts or beliefs. Christopher Hitchens, a popular journalist and author of several atheistic screeds, laments in his *God Is Not Great* about the detriment of Christian beliefs on society. Specifically, he writes vibrantly and passionately in an effort to convince people the Christian doctrine of the atonement is wicked and deplorable but without presenting a fair representation of the doctrine.

> [T]he idea of a *vicarious* atonement, of the sort that so much troubled even C. S. Lewis, is a further refinement of the ancient superstition. Once again we have a father demonstrating love by subjecting a son to death by torture, but this time the father is not trying to impress god. He *is* god, and he is trying to impress humans. Ask yourself the question: how moral is the following? I am told of a human sacrifice took place two thousand years ago, without my wishing it and in circumstances so ghastly that, had I been present and in possession of any influence, I would have been duty-bound to try and stop it. In consequence of this murder, my own manifold sins are forgiven me, and I may hope to enjoy everlasting life.[2]

Hitchens proceeds to tell us this foundational Christian doctrine robs humans of responsibility for their actions by teaching people they can impute their wrongdoings to someone else without consequences; they can find a scape-

2. Hitchens, *God is Not Great*, 209.

goat and be forgiven. He concludes by asserting the vicarious atonement is an "immoral theology."

Hitchens's tirade takes us back to our earlier discussion regarding atheistic presuppositions. Since Hitchens never demonstrated how, in his worldview, morality can even be a coherent concept, his subjective morality—for that is all the atheist can even claim to possess—lacks transcendent meaning and carries no *objective* truth for all people. This truly is the dilemma of atheism. Given the starting point of atheism, morality follows . . . how?

Why any nonbeliever condemns in others any act or belief contrary to his own convictions defies consistency since, according to his worldview, all that exists is matter in motion. And matter in motion, following the atheist's standard, is inherently without any morality. Since everything of this world can be explained purely in materialistic terms of physical processes, and because physical processes are categorically without morality, moral considerations have no logical place in his worldview. For instance, to say that dying for someone else is immoral makes as much sense as saying bile produced in the liver is morally wrong. To demonstrate Christianity is *objectively* evil, the atheist must first be prepared to tell us why it is evil. But this cannot be done following atheistic precepts. Atheism provides no coherent basis for rational condemnation in others when they misbehave or do vile things. Doug Wilson clearly exposes atheism's Achilles' heel in this exchange with Hitchens:

> Take the vilest atheist you ever heard of. Imagine yourself sitting at his bedside shortly before he passes away. He says, following Sinatra, "I did it my way." And then he adds, chuckling, "Got

> away with it too." In our thought experiment, the one rule is that you must say something to him, and whatever you say, it must flow directly from your shared atheism—and it must challenge the morality of his choices. What can you possibly say? He did get away with it. There is a great deal of injustice behind him, which he perpetrated, and no justice in front of him. You have no basis for saying anything to him other than to point to your own set of personal prejudices and preferences. You mention this to him, and he shrugs. "Tomayto, tomahto."[3]

Wilson's point is well-taken. Atheism cannot provide a rational basis for ethics or morality. For those, the atheist must rely upon the God of the Christian religion. The inconsistency of atheism leads us back to the only rational concept remaining: the supernatural is necessary in order to have an objective standard of morality. Hitchens and other like-minded atheists need the ethical imperatives of Christianity just to make their worldviews functional.

Stereotyping is a variation of name-calling and oversimplification, a combination of two previously mentioned faulty techniques. This approach takes one distinctive of a person, exaggerates it, pretends it is the only characteristic, and projects it on an entire group. Early twentieth-century American journalist H. L. Mencken took a dim view of fundamentalists, referring to them as "*homo boobiens*" and denouncing them as "uneducable." Fundamentalism, he argued frequently and derisively in his newspaper column, was only popular among the inferior orders of humanity.

---

3. Wilson, *Is Christianity Good for the World?*, 40.

To him, only the most gullible and uninformed members of society actually believe in a literal interpretation of Scripture regarding creation, miracles, and the accounts of Jesus. This negative impression of fundamentalism has changed little since the days of Mencken. Adhering to fundamentalism is, according to many of the so-called intellectuals, tantamount to endorsing the geocentric model or being a card-carrying member of the Flat Earth Society.

Another part of propaganda is the *glittering generality*. Propagandists express broad, far-reaching statements, usually ones with profound implications, but ignore the logical consequences of their rhetoric. "We'd be better off without religion in the world," they proclaim boldly. "The 'parties of God' have held countless billions of people over the course of thousands of years in their jaws, perpetrating deplorable acts toward each other and killing one another—all in the name of religion. Religions' lamentable record of murder, slavery, intolerance, bigotry, racism, colonialism, and so on and so forth is demonstrable proof organized religion is bad for the world. Violence and strife are the inevitable consequence of religion, Christianity included."

They bask in the simplicities of their platitudes, assuming religious tyranny is responsible for most, if not all, of the world's problems, while ignoring secular tyranny and failing to provide positive evidence that eradicating religion will contribute to societal progress. In fact, it could rightly be argued that since only one religious system is true—given that all religions and worldviews contradict each other at some point—every non-Christian belief system, including atheism, is an attempt to suppress the knowledge of God. All forms of aberrant religion are an attempt to subdue the knowledge of the one true God, so it is a categorical error to separate atheism from all other formalized religions and worldviews and pretend it is morally neutral.

On the question of morality, I fully understand why the atheist wishes to move beyond discussing the origin of ethical imperatives and ultimate meanings, because atheism provides absolutely nothing to the discussion—one way or the other. If Christianity is bad for the world, atheists cannot consistently point this out and provide a coherent definition of what "bad" even is, having no fixed way to define such an idea. Conversely, the atheist cannot even consistently argue religion is good for the world, since, again, there is no way to define good as a result of atheistic precepts.

More to the point, though, let's discuss the evils committed in the name of religion, specifically Christianity. What this individual is saying is that Christianity must be judged not solely upon those who profess the name of Christ Jesus and live accordingly to the truth and precepts found in sacred Scripture but also upon every single person who has been baptized in the name of Christ, even if they grow up to mock his very name and live a life contrary to the plain teaching of the Bible. In short, this atheist is arguing that those who excel in virtuous Christian living and those who repudiate gospel living are one and the same. I think this a curious way to argue and grossly unfair.

This does, however, bring us back full circle to the issue of ethics and morality. Namely, from where is an absolute, universal, and invariant system of mortality derived by which one can judge and make such dogmatic statements of truth—if not given by supernatural revelation? Whose worldview allows him to look at the Jewish ethnic cleansing of 1940s in Germany and make the true and dogmatic statement, "That was evil"? So who cares? Should the atheist even bother to think twice about the pogroms? If so, why—why *should* he even care? Moreover, upon what basis can the atheist objectively contend Hitler and his devotees should be held morally responsible for their actions if they are simply the consequence of the evolutionary process, masses of protoplasm controlled by indiscriminate brain impulses and nervous system patterns?

When we mull over the reality of imposing morality on this material universe and its inhabitants, we arrive right back where we started; we once again return to the idea of a transcendent and moral God. The atheist might like to spout glittering generalities about profound issues, but epistemological questions (how we know what we know) must first be established before meaningful dialogue can take place.

Slogans are another tool of the propagandist. Signs, billboards, and bumper stickers are all used to display messages. Often pithy and trite, these hackneyed phrases are displayed as truisms, yet are rarely challenged: "Gay Rights Equals Human Rights"; "The Right to Marry Belongs to Everyone"; "My Body Equals My Choice"; "All You Need is Love." Says who?

Transfer is a subtle technique that encourages us to transfer our emotions, positive or negative, about one source to another. Many of today's television sitcoms reveal portraits of independent female characters who epitomize today's intelligent, highly paid, and successful professional woman. These women may mock the importance of children having a father, instead choosing to bear a child alone and calling it simply another "lifestyle choice." Rather than depicting the realistic struggles single mothers face in a variety of race and class contexts, Hollywood represents a sanitized, well-to-do narrative of professional mothers who not only survive but also thrive financially and emotionally in father-less families. Other examples abound.

Among the menagerie of movie and television stars are a host of attractive homosexual characters who are depicted in an affectionate manner as being stylish, sophisticated, colorful, and warm so that the audience finds them funny and compelling. Hollywood social engineering presents us with a subliminal message that we should all be tolerant, loving, and accepting of who people are—regardless of their personal lifestyle choices—because, after all, they are some of the most endearing people. The goal is for us to transfer our positive feelings of homosexual characters we see on the screen regularly toward every day, mundane people who engage in this unnatural comportment.

Then there are the negative examples from Hollywood of Christians which, by themselves, could fill an entire book. These modern-day Neros seek to throw the Christian image to the beasts of secularism by portraying Christian characters as thoughtless, callous, hypocritical, and bigoted—in short, obnoxious and unlikable. Negative imaging isn't something new; it has been going on for decades. For a notable example of the backward Christian fundamentalist caricature, I think back to the 1960 film starring Spencer Tracy and Gene Kelly titled *Inherit the Wind*, a movie that portrays conservative Christians (i.e., creationists) as hostile to progress and intellectual views. Hollywood and the national media have since then and over the decades contributed in large part to the growing *anti* mischaracterization, resulting in the perception that Christian fundamentalists are against all sort of social "evils," such as smoking, playing cards, dancing, theater-going, drinking, and the like.

Testimonial is a technique that seeks to garner support for an idea because someone of steadfast repute speaks on behalf of the idea. Here is an example of the thinking: John is an important person. John wouldn't be so influential in Christendom unless he were better or more knowledgeable than we. Since he is preeminent in the studies of (fill in the blank) and is so important in the field, he must know better than we. Therefore, he understands what the Bible actually means regarding this issue and knows what he is talking about. We should believe him when he teaches on this subject. The idea is to encourage us to embrace (fill in the blank) uncritically because John is an influential figure and its chief spokesman. We are urged to endorse John's conviction because of our approval of John as an important figure in the larger rubric of Christian theology and not on the biblical merits of his argument. We might very well have sufficient grounds for embracing John's position, but being asked to accept John's conclusions solely because we find him impressive is unsatisfactory.

Statistics without context is another method used in propaganda. Statistics and polls are designed to shape public opinion, not reflect it—at least that's what the cynic might say. Propagandists give you plenty of numbers to "prove" their point but rarely provide insight into the background of the collection of their data. They often neglect to mention how many people were polled, the biases of those questioned, where the information came from, or how they gathered their statistics. Polling data can easily be skewed by pressing participants to opine on issues with which they have little familiarity, concern, or commitment. In effect, you can get 75 percent of respondents to support any position if you ask the right three people.

Arrant distortion or *stacking the deck* is another device of propaganda. Gamblers use a technique of stacking the deck in their favor by arranging cards in a particular order to ensure they will win. Propagandists selectively choose their information or "stack the deck" by ignoring certain facts to present a one-sided view. Since only a portion of the facts are displayed, we have no way of challenging them; hence, the deck is stacked against us, and we are helpless to respond.

For example, suppose someone says, "Early Christian writers plagiarized primitive pagan mythological figures from Horus and Attis to Dionysus and Mithra and concocted the Jesus story. Much of the material in the Christian Bible about Jesus's life, death, and resurrection is an amalgamation of folklore. Christianity, replete with its stories of creation, virgin birth, an only-begotten son, twelve disciples, resurrection from the dead, and so on, is nothing more than a plagiarized—albeit more sophisticated—version of ancient fables. Everyone living under the rule of the Roman Empire had a god who died and rose again; Christianity presented nothing unique with its claims."

What the propagandist fails to mention, however, is that the supposed ancient myths purportedly functioning as the bedrock of Christian claims had either insignificant or no presence in ancient Palestine at the genesis of the Christian movement. These beliefs simply were not in circulation or popular at or around the time of the birth of Christianity. The propagandists also blithely ignore the profound detestation shared by adherents of second Temple Judaism, and subsequently the early church, to the introduction of polytheistic deities into the culture. In reality, the early believers anathematized the very concept of pagan lore. They manifested, even in its earliest forms, an extremely exclusivistic faith system with a deep disregard for all forms of paganism. So to suggest intertestamental Jews and first-century Christians adopted the very concepts they despised, fashioned tenets from a host of competing and differing legends, and created a new religion—one whose foundational truths included at its core the concepts of honesty and absolute truth—is without historical or rational warrant.

Supposed parallels in Christianity with early pagan beliefs are not really parallels at all when critically considered. In truth, all these other gods of folklore differ from Christianity's claims about Jesus in weighty and crucial ways. For starters, God created the world *ex nihilo*, out of nothing, and not from preexisting matter, as every other religious claimant argues. God entered into his creation and became fully man while remaining fully God, not part divine, part man. The Spirit of God supernaturally impregnated Mary apart from divine intercourse with human beings, as the pagan myths taught. Jesus willingly gave up his life as a ransom for others; he was raised from the dead by the power of God, and did not merely rule the underworld like other deities.

And to submit one final example, Jesus serves as a mediator between God and humanity, making continual intercession for his people before a thrice-holy God—a wholly unique concept in any religion. When dispassionate observers examine all the evidence without a presuppositional bias against Christianity, they are forced to conclude that Christianity stands apart from every other religion or worldview and expresses a host of extraordinary and unique truth claims.

One final thought: As stated previously, there is nothing intrinsically harmful about propaganda, just as there is nothing inherently wrong with emotional appeals. What must be emphasized is the importance of recognizing when someone employs these tactics, because it is vital for us to understand how others are reasoning so we do not find ourselves trapped in a web of emotional appeals and allow ourselves to become victims of manipulation.

4

What Does That Have To Do with the Price of Tea In China?

"Enough about John Calvin," Linda barked. "I don't wanna hear any nonsense about a man who was responsible for the death of Michael Servetus. Why should I listen to the theological insights of a cold-blooded murderer?" she huffed. "Calvin was guilty of committing the worst kind of heresy imaginable, so none of us should esteem, let alone accept as valid, teachings Calvin perceives to be orthodoxy or scriptural. Quite frankly, I refuse to pollute the Kingdom by embracing the views of a man who tortured and killed people for merely disagreeing with him." With that, the conversation was over.

It was observing that scene unfold, a brilliant example of *ad hominem* argumentation, that finally convinced me to explore emotional appeals and logical fallacies in earnest. What I soon discovered was that Linda's technique is just one of many ways to avoid the essential issue or question. Like Linda, some resort to irrelevant appeals in an attempt to sway people's opinions with a specific attack that, though often persuasive and psychologically alluring, is really inconsequential to the facts of the issue.

Once we recognize irrelevant appeals, we will be better protected against forms of manipulation. Here, then, are types of irrelevance, many of them variations of previously mentioned appeals. We begin by returning to our example of Linda and her inappropriate dismissal of Calvin's theology *in toto*.

As mentioned before, the example that began this chapter is an *ad hominem* attack. It's easy to attack someone to circumvent an issue, criticizing the messenger instead of the message. This is precisely what an *ad hominen* attack is. A Latin phrase literally meaning "against the man," *ad hominem* argumentation is one method people use to evade a question or particular discussion by attacking the person making the argument instead of deconstructing the argument and pointing out why the conclusion does not logically follow from the stated premise(s).

For Linda, the theological truth claims of Calvin were overshadowed by lingering doubt in her mind of his innocence in the trial and execution of Michael Servetus, a condemned heretic.[1] A couple of points need to be marshaled in response to Linda's statement. First, even supposing Linda's accusations about Calvin were true, she doesn't really state objective reasons why his culpability necessarily renders his theological conclusions untrue. After predetermining his guilt, Linda reasons that his theology must necessarily be wrong, but this is not the proper way to argue. In truth, all fallen creatures undeniably have character flaws, and yes, some have even transgressed God's law with sinful activities analogous to Moses, David, and Paul, yet it does not natu-

1. For an historical account of Calvin's responsibility in the Servetus affair, see Bainton, *Hunted Heretic*.

rally follow that one's conclusions are always necessarily false because of this. While Calvin's views might not be factually correct, they are not automatically wrong merely due to deficiencies in his character. Spiritually bankrupt people can express true propositions. (For instance, the statement, "Jesus Christ was born in Bethlehem" would remain a true and valid proposition even if stated by an unbeliever.)

Second, one could properly ask, should all of Calvin's theology be called into question, including his particular views on monotheism, biblical inspiration and inerrancy, and the Trinity, or just certain doctrines? If the latter, upon what basis would one select which theological convictions should *not* be embraced and why? People who argue this way—and I have encountered plenty—inevitably find themselves on the horns of a troubling dilemma. These naysayers have no consistent basis for their choices other than arbitrarily selecting doctrines according to their personal preference. Someone can have character flaws, but that person's contention might still very well be true.

Every individual argument must be evaluated separately. Our responsibility is to figure out which arguments are valid and which ones are untrue, independent of other beliefs. Likewise, we should not automatically disregard all Calvin's arguments merely because he had convictions foreign to our own understanding of Scripture or possessed certain character flaws. If someone's doctrines are in fact erroneous, those doctrines must be exposed and proved to be spurious by Scripture and not jettisoned in favor of *a priori* prejudices.

Ad hominem attacks can take on the form of dismissing arguments solely based on perceived motives or viewpoints. Shane, a Roman Catholic, and Derrick, a Baptist, are engaged in a discussion concerning Martin Luther's understanding of the bondage of the human will and the Spirit's necessary enablement in the process of salvation. Shane replies, "Luther was a schismatic who rejected the authority of the pope and the Roman Catholic magisterium. He violated the clear teaching of Scripture by not submitting to God's ordained leadership. I don't think we should give much weight to your statement since Luther despised the Roman Catholic Church. He is obviously biased against the teachings of Rome and only seeks to separate himself from the church on as many doctrinal matters as possible."

One of Shane's problems is that his response does not account for the possibility that Luther accurately represented particular truths as expressed in God's Word—even if he rejects the so-called universal headship of the bishop of Rome. Shane's bias against those who reject the authority of the Roman magisterium is evident in his line of reasoning. He outright dismisses any argument that runs contrary to the teaching of Rome because of his presupposition (that is, no one outside the "infallible" teaching magisterium has the right to interpret Scripture). Shane subtly suggests that Luther articulated the doctrines of predestination and election out of spite for and in response to the Roman Catholic Church's rejection of Augustine's views on divine grace. So what's his tactic? It is simply to attribute Luther's conclusion to a personal vendetta against the medieval church while dismissing, without evidence, even the possibility that the Augustinian monk reached his conclusions from a thorough exegesis of Scripture. This is where Shane committed an error by imputing motives to Luther that are only assumed.

We should pause at this juncture to point out motives are not always pure, and we are in no way suggesting a person's motive does *not* influence his or her argument. We are merely saying a person's presupposition does not logically preclude the possibility of speaking truth, and it does not necessarily have bearing on the validity or strength of the argument presented.

Another variation of the *ad hominem* argument is *guilt by association*. A person is judged because of family members, friends, or associations, not because he has done anything particularly wrong. After attending a Sunday evening service in a small church in eastern Virginia, I headed for a table displaying an assortment of books, pamphlets, CDs, and general information about the church. The pastor ambled over to me as I was about to reach for one of the audio CDs.

"You don't want that," he cautioned sternly, his countenance clearly disapproving of my actions. "I meant to get rid of those CDs earlier. Somebody set them out this morning, and I forgot to dispose of them."

"What's wrong with them?" I asked unwittingly.

"What's wrong with them?" he scoffed, seemingly taken aback at my ignorance. "Young man, do you have any idea the conference that man has spoken at?

Well, no. And quite frankly, I wasn't overly concerned. I merely shrugged my shoulders, somewhat bewildered, and managed to respond, "No."

What I quickly learned was that this nationally recognized speaker, a man who had an unimpeachable reputation as an evangelical leader and someone with a strong emphasis on the family, spoke at an eschatological conference that, needless to say, was not the eschatological position shared by this particular gentleman. The pastor seemingly disapproved of the man's entire ministry because of a disagreement on a particular theological conviction. In reality, the speaker's end-times belief was irrelevant to the truth claims in the sermon recorded on the CD.

One point to glean from this experience is that we should focus primarily on what the person is saying and not necessarily his or her past associations. If this speaker's eschatological views were relevant to his message—aside from the fact that they were not—then that relevance would have to be established, not implied.

Another type of irrelevance is called the *tu quoque* argument, *tu quoque* being Latin for "you too." This particular method occurs when someone tries to dismiss a person's viewpoint because she is inconsistent on the very thing she condemns in others. Brian says to Melissa, "Don't criticize me for my lifestyle choices; you have made your fair share of unwise choices in the past." The issue here is Brian's personal actions, not Melissa's. Her behavior is a different issue altogether. Moreover, even supposing Melissa has been equally as reckless in her life as Brian is currently being, it does not excuse him of his destructive behavior. He is responsible for his own actions whether or not Melissa has engaged in the same sort of activity. Two wrongs, the saying goes, do not make a right.

The *counter-question* occurs when, instead of directly answering the first question posed, someone asks another question in reply. Jackson has this exchange with Robert:

"When Joseph states to his brothers in Genesis 50:20, 'You meant evil against me, but God meant it for good, to bring it about that many people should be kept alive, as they are today,' he is making a bold, declarative statement of compatibility between God's sovereign decree and the will and desire of humans. The point Joseph is stressing to his brothers is that God's overall intention was to save a great many people; it was his purpose to deliver Joseph from his brothers' hands and then into Potiphar's household in Egypt, finally to be cast into prison, where he would come into contact with Pharaoh, all for the exaltation and demonstration of his power and might. Joseph removes any doubt that his own personal experience in foreign captivity was merely God working a plan B. God had his intentions in the entire matter, and man had his. The Lord's intentions—purposes from before the events unfolded in time—were holy; man's intentions were wicked and deceitful. So tell me, if you can, where do you see libertarian human free will in light of this passage."

Robert replies, "Where isn't libertarian human freedom seen throughout Scripture? Besides, are you seriously arguing God decreed these events? If that's truly your position, wouldn't that make him the author of sin?"

Now these might be relevant questions at some point in the discussion, but not at this particular juncture. Before answering these distracters—for that's how they are being used—Robert is obliged to address Jackson's statement in a meaningful way. Otherwise, if he can't, and he has to resort to counter-questioning, he is possibly attempting to disguise the impotency of his position.

The counter-question is also commonly used, ironically, as an attack against Christians when discussing the merits of Islam. Consider the following example where a Christian addresses a troubling passage in the Quran that appears to legitimize the slaughter of non-Muslims: "After examining Sura 9:5 in context and scrutinizing the historical and scholarly Islamic documents related to this verse and *jihad*, there is, in my estimation, only one conclusion that can be drawn from Sura 9:5 when it commands Muslims to 'slay the idolaters wherever ye find them, and take them (captive), and besiege them, and prepare for them each ambush.' That is to say, this Quranic verse contains an offensive element in that it compels Muslims to destroy non-Muslims who do not repent. How can you seriously suggest orthodox Islam is a religion of peace?"

Not to be outdone, the non-Christian retorts, "The God of the Old Testament was no different when he directed the people of Israel to commit genocide against her enemies. So how can you argue Christianity is a religion of peace in light of such barbaric teachings, not to mention the atrocities committed during the Crusades?"

Again we can see the distracter in this brief dialogue. In this instance, the proponent of Islam attempts to justify or rationalize the teachings of Islam by suggesting they are similar to that of Christianity, though no evidence was offered in support of this position. The Christian can, in due course, rightly respond by pointing out man's sinfulness and guilt before God, divine wrath against willful violations of his law, fair punishment, and the propriety of human beings judging God, but all of those responses do not need to be given until the question has first been answered meaningfully.

Next there is the *irrelevant reason*. A group of young people are discussing the topic of wine during the celebration of the Lord's Supper. Gradually, the specific topic of alcohol becomes blurred; reasons are given against its usage, but none of these reasons properly address themselves to the topic under consideration but instead, to some other concern. "No, we definitely should not use fermented wine," claims one person. "The issue of Christians consuming alcohol, whether it's part of a religious ceremony or not, is too divisive for any church to insist on serving wine during communion."

Note that this person's argument is irrelevant. It has little to do with the propriety of serving wine. That some Christians object to alcohol *in toto* is a real but separate issue. If the Bible sets forth guidelines and a paradigm for the church, then that would unavoidably supersede personal feelings. The issue in this instance is not how do people feel about a particular issue—because, of course, that always leads to a myriad of unwarranted conclusions—but what does Scripture say? Some personal objection to alcohol is, in and of itself, no reason for adopting abstinence. Evidence mustered in support of or opposition to an issue must support the specific subject, not some related or tangential issue.

Following the irrelevant reason in a varying form is the *non sequitur*, another Latin phrase, this one meaning, "It does not follow." A *non sequitur* is a statement that does not logically flow from or is not clearly connected to any of the given premises. A cause-and-effect relationship is implied to exist when, in fact, there is no rational way to connect what was previously said and the stated conclusion.

Observe the following statement a Roman Catholic addresses to a Protestant: "There are today, according to a recent study, in excess of thirty thousand distinct Protestant denominations around the world, each of which claims to adhere to the doctrine of *sola Scriptura*, the Bible alone. Yet no two denominations agree on what precisely the Bible actually teaches. The doctrine of Scripture alone is unstable and leads to a myriad of conflicting, erroneous, and oftentimes spiritually pernicious human traditions that lead people away from the truth expounded by Christ and the apostles. Therefore, without the infallible teaching magisterium of the Roman Catholic Church consisting of the pope and bishops, we would never definitely know for certain things necessary to be known, believed, and observed for salvation. That is, apart from the living magisterium of the church, we would not know what books belong to the canon of Scripture, the doctrine that the persons of the Trinity are *homoousios*, that in Christ there are two wills, the hypostatic union, and a host of other fundamental issues that bear directly upon the core of the Christian faith."

The conclusion offered here is a lucid example of a *non sequitur*. The assumption that the Roman Catholic magisterium is required for infallible knowledge about biblical truths has nothing to do with the Reformation doctrine of *sola Scriptura*. Even if we were to grant for the sake of argument that there are deficiencies in the understanding of the Bible alone to render it impractical and self-contradictory, the conclusion offered by the Roman Catholic does not naturally follow. We are never given reasons why the Roman magisterium—over against any of the numerous other so-called infallible interpreters of Scripture—logically emerges as the true authority for the church purely as a consequence of dismantling the Protestant tenant of the Bible alone.

In reality, this line of argumentation is woefully insufficient and does not even begin to address how the established church and her traditions, like all Scripture, are "God-breathed" (2 Tim 3:16), which is the necessary requirement for an authoritative voice in the church. Until she can meaningfully demonstrate her traditions are of the same nature as the Bible, there can be no higher authority for the church than what ultimately originates from the very breath of God. In short, the Roman Catholic response never gave us any compelling reason why the Roman magisterium should be embraced unceasingly and unquestioningly. Instead, it offers us a hollow argument with a conclusion not supported by the premise—a *non sequitur*.

Another prime example of a *non sequitur* is when some insist the doctrine of predestination undermines the urgency and gospel need for evangelism. "Since God chose the elect in eternity past, and since he alone grants repentance and faith necessary in order to embrace him, and since he has chosen all who will, indeed, come to him, then it follows that our efforts are useless in the process of salvation. After all, if someone is preordained unto salvation, he or she will eventually get saved—*que sera sera*, whatever will be, will be."

Again we see an incongruous conclusion. There is nothing concerning the biblical doctrine that necessarily leads to devaluing evangelism or precludes any zeal for the Great Commission. Believers are very much involved in bringing salvation to the world. God ordains not only the redemption of his people but also the means unto salvation. He has specifically chosen the "folly of preaching" to communicate the gospel message, the instrument by which unsaved people are made conscious of their own desperate plight, repent, and embrace Christ as Lord and Savior. Evangelism, therefore, according to the Bible, is the responsibility of all believers. What's more, understanding the promises of God provides greater certainty that evangelistic efforts will yield tangible results. It demonstrates that our evangelistic efforts will not be futile because of God's promises. In fact, predestination should serve as a greater impetus for missions, just as history has so amply demonstrated with the notable Calvinistic missionaries in generations past.

An additional way in which people attempt to persuade others is by adding *irrelevant details* into the conversation, introducing prejudicial intent that is unwarranted and immaterial to the matter. Here is an excerpt from a column that appeared in the *New York Times'* Opinion section, whose author was guilty of using irrelevant details:

> Chief Justice Roy Moore of Alabama has built his career on demagoguing about the Ten Commandments. As an obscure state court judge, he posted the commandments on a wall behind his bench, and used the controversy over the display to get himself elected to Alabama's highest judicial post. Once in office he installed a two-and-a-half-ton Ten Commandments monument in the rotunda of the main judicial building in Montgomery. When two federal courts ordered him to remove it, he resisted, claiming

> that he is an independent constitutional officer. Now that he is facing a court deadline of Aug. 20, he is milking the drama for all it's worth. He says he will announce tomorrow whether he will comply.
>
> There is a very serious principle at risk in Justice Moore's grandstanding. The federal Constitution applies to the states, and the federal courts are its ultimate interpreter. Justice Moore's desire to ignore the Constitution's mandates on the separation of church and state has an uncomfortable resemblance to the arguments Gov. George Wallace made when he mounted his stand in the 'schoolhouse door' to block blacks from enrolling at the University of Alabama.[2]

This skittish invective is full of spicy, colorful language that could be summarized, albeit unfairly, in this way: Judge Roy Moore has a greater concern to promote his primitive beliefs regarding a prominent display of the Ten Commandments than he does to uphold the Constitution. It would be difficult for anyone to get through the first paragraph without forming a negative opinion of the judge. From the opening sentence, the author fires off a torrent of baseless accusations, manipulating our attitude toward Judge Moore instead of presenting us with facts and letting us formulate our own opinion.

Words such as "demagoguing," "obscure," and "milking," are irrelevant—and the author's comments are deliberately selective about what information was portrayed. The writer could have chosen other statements to present

---

2. Anonymous, "Justice Roy Moore's Lawless Battle," 1–13.

the facts of the case or about the judge's lifelong record but instead introduced an entirely different impression of the judge. Notice how the article was introduced by referring to Judge Moore as a civil servant who built an entire career "demagoguing about the Ten Commandments" and informed us he is unquestionably an "obscure state court judge." Of course, we are never given any objective measure by which to define these things; instead, we are told they are so. It's almost as if the writer is telling us, "I said it, so believe it, and don't question me." Note also the gratuitous injection of negative adjectives, words that give the impression the judge is a radical, right-wing fanatic who "clings to his religion and guns" and manipulates the law to suit his desires. It is insincere and irresponsible to write in this manner with this tone, inserting statements without providing tangible evidence and pretending they are objective facts.

The *appeal to force* is used when someone applies tension to get someone else to do what he or she wants, making him fear the consequences. The 1520 papal bull entitled *Exsurge Domine*, promulgated by Pope Leo X in response to Martin Luther's *Ninety-Five Theses*, is one such example. The pope rejected Luther's writing as "heretical, scandalous, false, offensive to pious ears or seductive of simple minds," and warned that any person, regardless of sex, who read Luther's *Theses* or any other writings immediately would fall under the "penalty of an automatic major excommunication."[3] The use of pressure or force might serve its purpose on occasion, but it does not necessarily constitute sound reasoning for doing something. Being placed

3. For a discussion of the papal bull, see Lepicier, *Indulgences*, 325–31.

under the threat of ecclesiastical excommunication is not a sufficient reason for conforming; it is merely a description of what will happen if one does not cooperate. This argument, in effect, says that the person who wields the greatest power is right. *Submit or face the eternal consequences*, the Roman Church threatened the denizens of Europe without effectively stating why.

Still another irrelevant appeal is the *appeal to ignorance*. This technique can take several forms—namely, that since one cannot prove one's claim, it must therefore be false; and, conversely, since one cannot disprove what the other person is saying, he or she must necessarily be correct. For instance, "No one has ever been able to prove that God exists; therefore, he does not exist." Equally, "You cannot cite for me one clear example of papal fallibility. In fact, there is no such example from history regarding any pope erring on matters of faith or practice; therefore, your claim against the infallibility of the pope is unsubstantiated and false."

Granted, one should not accept a premise in the absence of proof, but what is important to remember is that the absence of proof does not nullify the premise or automatically mean something is wrong. It merely means that no evidence has yet been proffered in support of the stated premise, not that no evidence exists. My not being able to cite specific examples of papal corruption does not mean the bishop of Rome, even when supposedly speaking *ex cathedra* ("from the chair" in an official capacity), expresses infallible statements on matters of faith and morals; it simply means I can't remember any particular instance from history that, in my opinion, is fallible. Perhaps I read about

Pope Honorius teaching the monothelite heresy, Popes Victor (189–198), Zephyrinus (198–217), and Callistus (217–222), all of whom embraced the anti-Trinitarian heresy of Sabellianism, Liberius's endorsement of Arianism with his signing of an official decree in 358, Felix II's overt Arianism, and Sixtus V's erroneous Latin edition of the Bible, to name just a few, without having committed the specific details of these examples to memory.[4]

Put simply, the paucity of illustrations or lack of evidence does not disprove a claim, nor does it automatically mean a position is false. The weakness might very well be with me and not with my claim.

Next we come to the *vague appeal to authority*. This is a specious tactic that might give the impression of weightiness but often turns out to be inadequate upon further investigation. "The overwhelming consensus is . . ." "The majority of biblical scholars say . . ." "They are now sure that . . ." When statements such as these go unchallenged, they appear to lend credence to the proposition that follows. The statements, however, should not automatically be accepted simply because some nebulous group is cited in support of a particular position. Facts and evidence must accompany assertions if they are to be believed.

A priori is also a type of appeal to authority. When someone argues *a priori*—a Latin phrase meaning "formed or conceived beforehand"—one is arguing from theory. One regards what he thinks should be true as a perfectly valid reason for accepting it as if it were true. Imagine hearing

---

4. For further discussion regarding papal infallibility, see Rowell, *Papal Infallibility*. Also, Pusey, *Is Healthful Reunion Impossible?* and Murrell, *So You Want to Become a Roman Catholic?*

this conversation that, I believe, is reflective of the mood and thinking among many Arminian evangelicals. Bryson, for example, states, "God would not command us to do something that we, in and of ourselves, are not capable of doing. So when God commands the entire world to repent, each and every person has that innate ability to conform to his divine directive apart from any additional measure of grace." The fact is Bryson is not really sure if this is how God and man both operate; he merely hopes or suspects that this is the case since there is nothing in God's revelatory statements about repentance and obedience that naturally leads to the conclusion that all humanity has a sense of autonomy and ability.

What Bryson overlooked in this instance is that when God gave the law to Israel, it was to uncover what people *cannot* do, not what people *can* do. That is to say, God gave his law to expose iniquities and to increase humanity's guilt so that no person standing before the righteous judge could declare his own righteousness. Instead, it was given to reveal to every person his or her own shortcomings so every mouth would be stopped and no person would stand before his or her maker and declare, "I'm not *that* bad." I'm reminded of the famous story about Martin Luther in which he is reported to have told his chief theological nemesis, Erasmus, "When you finish writing out all his commands and exhortations, all the deeds humans must accomplish, I will write 'Romans 3:20' over the top of it all." Luther's point is well-taken. Likewise, it is folly to attempt to demonstrate libertarian free will from the Old Testament Law when it was given to demonstrate mankind's sinfulness and perpetual inclination toward evil.

What's more, it seems as if Bryson hasn't considered the possibility that God had a greater reason for commanding his creation to do something, even knowing they no longer had the innate ability to obey because of the effects of the fall of our first parents. Thus, it could very well be that God wanted to show us our inability to perform his commands. With that in mind, nothing can be deduced about libertarian free will absent a direct correlation in Scripture. Unless Bryson has concrete evidence to support his claim, he cannot know with any degree of certainty about mankind's ability to obey God apart from divine intervention from a few commands in Scripture. Even if it is an educated guess based upon experience and his philosophical understanding of God's character, it is still only a guess.

Sacred cows are the ideas, notions, principles, or beliefs that are held sacrosanct. When someone refuses to allow you to question a particular statement and accuses you of challenging one of these deep-seated ideas, he is using the irrelevant appeal to the sacred cow. "What do you mean by criticizing the Marian dogmas? Are you against the authority Christ invested in the pope, the Vicar of Christ who rules as supreme head of his external church?" Or perhaps this one: "When you criticize the preacher, you criticize all that he stands for: ecclesiastical authority and biblical Christianity." Your assault on a particular issue or statement does not mean that you are attacking the person or the ideals for which he stands, only a specific proclamation.

Sometimes *jargon* is used to give the impression of profound introspection or authority. We may transfer our positive reaction to people who use lofty statements or phrases to communicate their thoughts, making us accept their views. The postmodernist boasts that objective truth is merely an illusion: "The revolutionary era is upon us—an era liberated from oppressive strictures of the past but at the same time disquieted by its expectations for the future, discovering the critical, strategic, and rhetorical practices employing concepts such as difference, repetition, the trace, the simulacrum, and hyper-reality destabilize antiquated concepts such as presence, identity, historical progress, epistemic certainty, and the univocity of meaning." This may sound impressive to some people, but the words are vacuous and convey utter nonsense. Always be on the lookout for people who use pompous jargon to communicate their thoughts; it might be a good indicator that they are trying to camouflage complete nonsense. You can put lipstick on a pig, but it's still a pig—meaning, you can dress something up, but it doesn't change what it is.

This chapter focused on a number of irrelevant appeals that are commonly found in everyday discourse. If you listen closely, perhaps you will discover even your own encounters with others are filled with some of these irrelevant tactics. If so, it is imperative not to let these techniques falsely persuade you. Furthermore, we must all ensure we are fairly and accurately dealing with the thrust of an issue and not become bogged down in the litany of superfluous details that have no bearing on the matter. Always make sure to let the facts get in the way, for truth is worth discovering—wherever it may lead.

5

A Straw Man Never Fights Back

I RECENTLY came across a delightful *Peanuts* comic strip. The cartoon shows Charlie Brown and his little sister, Sally, outside, the latter gazing heavenward in awe while the former stands with a quizzical look on his face, wondering what all the fuss is about. "Isn't the sky a beautiful blue today?" Sally asks Charlie pensively. "'Beautiful' is only a relative term . . ." Charlie replies. "As a matter of fact, so is 'blue' . . . Color, you know, is beautiful only when it is good color . . . Of course, then you come up against the question, just what is good color?"

"Oh, good grief!" Sally can only walk away, irritated with Charlie's lack of appreciation and his bland microanalysis.

Many people today, much like Charlie, are guilty of committing the same error of thinking—the error of diversion. Sometimes intentional and other times by accident, diversion can be useful when you are backed into a metaphorical corner, when you feel as if you are losing an argument or debate, or when you feel adrift or uncomfortable with the direction of the conversation. You can do what Charlie did, since he didn't care for the topic, and change the dialogue into a formal analysis. Alternatively, the error of diversion can be used against you if you're not alert or aware of what is happening. It is just as easy to be distracted by this error as it is to distract others from the original topic. Once emotion or irrelevant details are mixed into the equation and go unopposed, then the discussion will inevitably get sidetracked with unfortunate discussions that become the main focus rather than what the speaker originally intended.

Many of the preceding examples from earlier chapters can, and often do, lead to diversions. Distracters only serve to complicate an issue and muddy the water so that the audience is oftentimes led into a wilderness of confusion and ambiguity. Avoiding this pitfall is essential in nurturing a reasoned argument or discussion. The chief aim of this chapter is to supply you with enough examples to caution you against this very approach. If you recognize when someone is using diversionary tactics, you might be able to avoid being deceived.

We will discuss the two prime sources of diversion: the *red herring* and the straw man argument, two ubiquitous methods that are present in almost any debate. We will begin with the red herring.

THE RED HERRING

To train young scent-hounds, a dead fish—more specifically, a red herring—would be dragged across the canines' paths, causing the hounds to pursue a different direction. The odious scent of the fish readily distracted the dogs from their original goal until they were taught to remain focused. Eventually, the hounds learned to ignore the scent of the red herring, as alluring as it was, and were able to follow the original scent rather than the more powerful, pungent odor. The idiomatic expression of a *red herring*, then, is an attempt by someone to get us to chase after disturbances during the course of a discussion, distracting us from the real issue. As one might expect, red herrings can turn a discussion into an unobstructed dialogue of irrelevant facts and might result in a litany of superfluous exchanges that never return to the original discussion.

Observe the subtleties of the following conversation, and see if you can pick out the distracters from the relevant facts. James and Patrick are deep into a discussion about the Bible and its role in the church and in the life of the believer today.

> JAMES: I'm not saying we don't use or cherish creeds, confessions, or tradition. All I'm pointing out is that all of those are subordinate to the Bible and cannot hold a person's conscience captive like the Bible can because they are not divinely inspired. Remember, historic Protestantism affirms the Scripture to be the only written revelation from God. The apostle Paul informs us *all* Scripture is "breathed out by God" (2 Tim 3:16), distinguishing it from all other human writings. As a consequence of being a divine product from the mouth of God, the Bible is infallible and inerrant in all it affirms; the testimony of Scripture is true (unlike some creeds, confessions, and traditions that can, and have, erred). Moreover, since the Bible's very nature is God-breathed, it is sufficient, containing everything

necessary for salvation and the duties required of us. Just think about this for a moment.

The canonical Scripture is the voice of God; it is his divine artifact of revelation preserved for his bride, the church. For this very reason, we submit our thoughts and our moral standards to the Bible *alone*, since, again I repeat, it is God-breathed. Unless you can demonstrate to me your ecclesiastical tradition is likewise God-breathed, by its very definition, your extra-canonical tradition necessarily remains subordinate and inferior to the Word of God. In other words, the Bible stands as *the* ultimate authority, because it is *theopneustos*, God-breathed, and therefore embodies the very speaking of God, and must, logically speaking, be of the highest authority, allowing for no equals.

PATRICK: Yes, I understand implicitly where you are coming from because I too once believed as you. I remember well how I struggled against the church's rich pedigree traced from Peter throughout the ages. I remember how I put up a vigorous fight when I began studying the church's teachings . . .

JAMES: The Roman church's teaching. You said, "The church's teachings," but I think you meant, "The Roman church's teachings." Since the bishop of Rome alleges absolute supremacy in faith and morals over the entire visible church, one can hardly suggest the church at Rome is by any stretch of the imagination *catholic* or *universal*. My point is simply that we need to be specific and say your beliefs are in line with the

Roman Catholic Church and not merely using the amorphous label "church."

PATRICK: Well, be that as it may, Christ established only one church, so I'm not going to harangue over semantics. Getting back to my point, let's just say that I was a staunch anti-Catholic when I was a professing evangelical, so it took a tremendous amount of time before I came to see the truth of the Catholic church's teaching.

JAMES: I'm sorry, but you said you were an anti-Catholic? What exactly do you mean by using that moniker? Are you saying you wrote books against the Roman Catholic Church, engaged in formal debates, preached against the Roman magisterium—that sort of thing?

PATRICK: Well, no, I didn't do any of those things.

JAMES: Did you at least write tracts, articles, or blogs against the Catholic church?

PATRICK: No, I didn't do any of that either.

JAMES: Okay, I see. Then what you really mean by "anti-Catholic" is that you, like most convinced evangelical Protestants, rejected Rome's claims of absolute authority in all matters pertaining to faith and practice, right?

PATRICK: My point is simply that I understand your perspective; I get where you are coming from because I have walked down the same path on my journey toward the true church, coming into communion with Christ's church built upon Peter. You see, it was my inability to coherently define and defend, from a biblical and logical standpoint, the doctrine of *sola scriptura* that truly caused me to make the decision to submit to the infallible teaching magisterium of Rome.

JAMES: Before we proceed any further on this topic of the Bible alone, I think it would be best if we refrained from the personal testimonies. I could also point out many who once embraced Roman Catholicism, only to reject it later in life. We are not seeking to explore your personal testimony, as curious as it might be to hear, but please proceed.

PATRICK: Well, in any case, given the fact that there are in excess of thirty thousand different denominations, with new ones springing up every week due to *sola scriptura*, you can easily see how such an unbiblical doctrine leads to anarchy. Christ did not establish his church only to see it fragmented and splintered the way Protestants have been attacking it since the Reformation period. Luther, Calvin, Zwingli, and your other "heroes" of the faith are nothing more than schismatics who rejected the authority Christ established. Any way you slice it, going by the *Bible alone* is a recipe for disaster and should be rejected because of the pernicious effect it has had on the established true church.[1]

Whatever happened to the original view of the Bible being the sole rule of faith for the church because of its divine nature? As much as James has tried to keep the topic focused, he and Patrick have deviated from their original

---

1. For a helpful dialogue regarding *sola scriptura*, see White, "Sola Scriptura in Dialogue."

topic, chasing down rabbit trails that are irrelevant to the thesis of the discussion. They followed the red herring of labels and then followed another red herring of the nature of the church and personal conversion experiences until Patrick ended on the distraction of alleged division in the evangelical community purportedly wrought while seeking to adhere to the principle of *sola scriptura*.[2] James could justly respond (as Plato rightly noted centuries earlier in his *The Laws* regarding this type of faulty reasoning) that an abuse of an authority—in this instance, the Bible—does not render the Protestant principle untrue. Their chances of having a focused discussion relating to the biblical evidence for the church functioning with the Bible alone as the *ultimate* authority are slim.

Another deflecting tactic common in dialogue is the use of sarcasm, humor, ridicule, parody, or innuendo, which can lead to a host of diversions. These measures can alter the course of the discussion and promote an entirely new line of thought. A person becomes insulted and tries to save face, turning his remarks into a rebuttal period instead of focusing on the direct issue.

Then there is the witty remark, a quip intended to make people laugh, and while the audience is laughing, capitalize on the moment to turn the tide of the discussion. In a debate with a Roman Catholic, one particular evangeli-

---

2. Ironically, Roman Catholic apologists seemingly forget Protestantism derived from within the confines of the so-called united Roman Catholic structure. It is grossly disingenuous to argue Protestantism causes division and strife when the "unified" church produced the schism that split the Western church in the sixteenth century. If the same argument one advances could be used against one's own position, then it is a failed argument.

cal apologist argued on the basis of Scripture that salvation is entirely an act of a gracious God who grants saving grace to his people on the basis of his grace, through faith, and *not* on the basis of human works or deeds. When the opportunity for a rebuttal came, the Roman Catholic debater quickly retorted, "There is only one time in all of Scripture where the phrase '*faith alone*' is found, and it is preceded by the words '*not by*'!" Such simple platitudes might get a ringing endorsement from some faithful followers who are not concerned with meaningful and exegetical studies, but the intent of the debate was a biblical analysis of Paul's view of justification and not a misreading from the book of James—a book that focuses on proper Christian attitudes following a profession of faith. It is not intended as a discourse on how a sinner is made right before God in the Pauline sense of the word. Nonetheless, the Roman Catholic's witty remark diverted the thrust of the debate into an entirely new, albeit distorted, direction.

Yet another example of a red herring occurs when one puts one's opponent on the defensive by using highly technical jargon or esoteric facts designed to make one's opponent appear ignorant and thus, wrong. One particular evangelical points out to a colleague that we do not need the repetition of the Mass because, as the apostle noted, "We have peace with God"; we have a subjective and an objective cessation of hostilities solely because of the finished work of Christ at Calvary. In other words, there is nothing left to accomplish; there is no more work to be fulfilled. We, as believers in Christ, have accord and are relationally at peace with God.

The Roman Catholic replies, "Are you aware that Romans 5:1 is a clear example of itacism and that while the manuscript tradition attests both the indicative for *echomen* ("we have") and the horatory subjunctive *echōmen* ("let us have"), the subjunctive form has the stronger external attestation among the two most important Alexandrian uncials?"[3] Even supposing this statement is true, it would not invalidate the idea that Christians have an abiding peace with God and do not need to be constantly weary of the need to perpetually do good deeds to placate his wrath (e.g., attend the Mass daily to wash away our sins). The latter translation in no exegetical manner invalidates the former, yet how many people can see through the smokescreen of the technical language?

Consider the diversion that occurs when a person makes a petty objection. You make a frivolous mistake, and your opponent pounces on that single peccadillo, even if the mistake has no bearing on the point you are trying to make. You might be able to overcome it with great aplomb, but some may become distracted, flustered, or even lose credibility in the sight of others. People might falsely assume your testimony is no longer credible, even if your points are valid.

For instance, you and a friend are discussing the merits of the Marian dogmas and are ruminating upon the matter of the Assumption of Mary. You, arguing against the Catholic dogma, make the point that Mary's *ascension* into heaven was first propounded by Gnostic literature several centuries after Christ, citing copious examples of Patristic writers who discussed Mary at great lengths but failed to

3. See discussion in Moo, *The Epistle to the Romans*, 295–96.

see or to mention this seemingly obvious and crucial doctrine. Your friend then jumps upon your misstatement, claiming Mary did not *ascend* to heaven but was *assumed*, and she makes a monumental fuss over your not using the correct term. Granted, it is true the dogma speaks of assumption and not ascension; however, the fact remains that the error in terminology has no bearing upon the reality of the shortage of biblical or ecclesiastical evidence to justify the dogma. Your misstatement does not invalidate the soundness of your argument, though the tenacious rebuttal of that single point, and that point alone, demonstrates the weakness of your friend's position.

Finally, a person might try to create diversion and confusion by feigning ignorance on a certain matter, claiming he doesn't understand what was said when in fact he is using this technique to make his opponent stumble over a convoluted matter in hopes of making him look foolish. He may play dumb and try to get the person to re-explain his position in hopes that the speaker muddles the issue—and the audience in the process. On the other hand, he might just make himself appear foolish or ignorant in the process.

THE STRAW MAN

The fallacy of the straw man is to take something your opponent has said and exaggerate or change it radically while replacing it with a superficial similarity to make it easier to refute. When you do this, you are creating a straw man—an easier representation of your opponent's statements.

The following comments stem from a short essay in response to Calvinistic soteriology:

> Calvinism is nothing more than hybrid paganism. The "doctrines of grace," as some like to refer to the five points of Calvinism are without reservation fatalistic, predetermination reaching its uttermost extremity. Here the imagination is strained utterly as the Calvinistic "gospel" tells multiplied millions and billions of Adam's fallen race: "You have no hope because God has abandoned you." This is the gospel? This is good news? And to adapt a well-known title: *What Love Is This?!* This Calvinistic message is as cold as a snake and as deceitful as Lucifer. So, the revolt against God continues. Many able theologians have powerfully described the fallacies of the Calvinistic system as "theologically inconsistent, philosophically insufficient, and morally repugnant." I couldn't have said it better myself—Amen![4]

Of course, in reality a vast chasm exists between pagan notions of fatalism—a mechanistic determinism antithetical to the notion of a personal, immanent God—and the classical Protestant view of a sovereign God who rules over his creation, bringing his will to fruition, unfolding his intended purposes to accomplish a plan that ultimately brings him the most glory and honor. The former is grossly depersonalized; the latter is intimately connected with a loving God who is actively involved with his creation by working out *all things* for good (Rom 8:28–30). And that's just the

---

4. As cited in Hendryx, "Arminian Suicidal Tendencies," 8–10.

first of many egregious misrepresentations of the theological framework known as Calvinism—without even delving into the actual Calvinistic concept of humanity, sin, the wrath of God, the nature of the will, the grace of God, and so on, all aspects that were grossly distorted in the above diatribe.

Sometimes a straw man can be created by extending the opponent's ideas even when there is no rational warrant for doing so. "You are in support of X; then, next, you'll be advocating Y, and then Z." Your opponent attempts to show how fantastically ridiculous Z really is, trying to make your position look foolish. Of course, you never even remotely mentioned anything connected with Z, but the audience has all but forgotten that little tidbit at this point. "Now that it is clear to everyone that you affirm the Calvinistic doctrines of salvation, you cannot escape the inevitable conclusion that predestination and election destroy all zeal for an upright life. There is no rational reason why any person should ever concern himself or herself with living righteously since the elect will inevitably get saved at some point before their death and the rest will perish anyway. In fact, exhortations to righteous living seem rather inept and a sheer waste of time if your understanding were true. So in the final analysis, these doctrines are a recipe for disaster—a perfect storm for the death knell of evangelism."

This statement is a glaring oversimplification of the doctrines of grace. There is nothing inherent in the doctrines of predestination and election that would require one to assume morality is inconsequential to human agency or that God could not ordain the means as well as the ends to achieve his divine plan.

Still another way to create a straw man is by attacking an example. In the process of proving his point, your opponent constructs an illustration or analogy he thinks will dismantle your position. After articulating this example, he begins to dismantle the illustration by pointing out problems, hoping to discredit or weaken your position. If you are not careful, though, you might just let him incorrectly associate his anemic example with your position, being unaware of just how poor the comparison is. Consider, for example, the following analogy used by one prominent evangelical apologist in one of his books illustrating, from his perspective, the God of Calvinism:

> There was a farmer who owned a pond. He did not want anyone to go swimming in it. He built a fence around the pond and posted a sign that said: NO SWIMMING ALLOWED.
>
> One day three boys came upon the pond. They saw the sign, but decided to go swimming anyway. They climbed the fence, and jumped into the pond. After jumping in, they realized that there was no way to get out. They began to drown.
>
> The farmer came to the pond, and saw that the three boys were drowning. He said to the boys, "Didn't you see the sign? You have broken the rules. But I am a kind and loving farmer, so I will let one of you out." The farmer then proceeded to throw a rope to one of the boys, and pulled him to shore. Then the farmer folded his arms and watched the other two boys drown.[5]

---

5. Geisler, *Chosen But Free*, 50. Geisler advances a similar argument in a chapter entitled "God Knows All Things," in *Predestination and Free Will*, 69–70.

While some might initially be drawn to the analogy offered, the illustration flounders on many different levels. First of all, it should be pointed out that in this example, God is compared to a farmer (as if this accurately portrays the stature of Almighty God), while sinful humanity—those whom the Bible says naturally hate God and follow after the devil—are likened to a few adventuresome boys who merely want to have fun by sneaking past a signpost to swim in a pond. Perhaps a better analogy of fallen humanity would be to have the boys not only swimming but also robbing, plundering, and pillaging the man's house. The original illustration, if the analogy were truly a fair comparison, would depict the boys as reveling in their machinations and encouraging others to participate too, creating all sorts of mayhem and violence even after this gracious man has housed, clothed, fed, and protected the boys, even in spite of their long track record of hatred and rebellion toward him.

Then, after having established the true plight of the adolescents, we would need to point out that if the gentleman made such an attempt to save to boys from drowning, there's a strong chance they would mock him for his efforts. In fact, they may obstinately and freely refuse to leave the waters; they would curse the man, spit in his face if they could, and would remain where they were—in the midst of chaos and certain death. Given the opportunity, they then attempt to drown the man, pulling him under and destroying him if it were possible.

For all intents and purposes, the man should have ensured justice was served and seen to it that the perpetrators died in the pond for their egregious atrocities against the man and his family. Displaying unmerited favor, the man instead shows an inimitable love for those who hated him and wanted to destroy him; he sends his only son into the water to rescue the boys who, by all accounts, had no desire to be saved. They did not ask for redemption; they tried to flee from it. Yet the father sends his son to save the boys and in the process, loses his only son after saving the rebels. In light of numerous atrocities and opprobrious behavior, the man is completely justified in saving only some of the boys. He saves them freely by his grace, even though, in actuality, they all do not merit his kindness and deserve to perish. This, I submit, would have been a truer analogy of the Calvinistic understanding of humanity and the nature of God.[6]

Analogies must be carefully crafted to ensure they are proper representations of the position they are attempting to challenge. Otherwise they are nothing more than straw men arguments.

Of course, it is easy to become distracted by irrelevant details, chasing after nonessential issues, following red herrings, and discussing straw man arguments that do not focus fully and fairly on the specific issue. We must always be on the alert so as *not* to confuse what appears to be relevant with what actually is relevant. Cultivating the ability to discern distractions amidst all the rhetoric will equip one to stay focused on the real issue and prevail in stating one's case.

---

6. For an expanded rebuttal to Norm Geisler's fable, see Storms, "How Can God Be Loving? Also White, *The Potter's Freedom*.

6

Why Assumptions and Incorrect Inferences Get Us into Trouble

A FAMOUS story is told about Harry Houdini making a "jail break" tour of European cities in the first half of the twentieth century. During one of his displays to showcase his dazzling talents, the great Houdini, escape artist extraordinaire, found himself center stage during a very ironic event. After having been stripped and searched to ensure no key or lock-picking device was being concealed, Houdini, as part of his magical act, was manacled in a Scottish town jail to prove no lock, however prized it was thought to be, could constrain the "elusive American." The old turnkey shut Houdini in a cell and departed—yet it was not long after the door shut that Houdini, just as he had done many times before, was able to free himself from his shackles effortlessly. Once free, he began working on the cell lock, but it proved to be a tougher challenge than he anticipated. He worked and worked, but despite all his efforts, the lock would not cooperate; it would not open. Houdini was running out of time.

Finally, exhausted, desperate, racing against the clock, and out of ideas, he leaned against the door—and it instantly swung open so unexpectedly that he nearly tumbled headlong down the corridor. In all the excitement of the famous visitor, the turnkey had failed to lock the door.

Similar to Houdini when he saw the door close, we all are prone to make false assumptions. At times we take things for granted or accept ideas or concepts without sufficient justification for doing so. This is not necessarily a bad thing, but we should be keenly aware of our assumptions (and the assumption of others).

Let's embark on a little thought experiment, shall we? Identify the letter that is different from the others in the following: *a*, *b*, *c*, *q*.

Which one did you choose and why? Take a minute if you need.

Now let's evaluate the possibilities.

The letter *a* is different from the others in that it is the only vowel listed, it is the first letter of the alphabet, and it is the only one with a short stem on the right.

The letter *b* is different in that it is the only letter with a stem on the left side and a stem pointing upward.

The letter *c* is different in that it is the only letter with a break in its circle instead of a stem, and it is comprised of only one part.

Finally, the letter *q* is different in that it is the only one that does not follow alphabetically, and its stem points downward.

It would be hard to make the case that any one of the letters is more different than the others, yet you probably picked a single letter—possibly *q*—because you made certain assumptions or inferences. When we do this in life, it can ultimately lead to confusion or to wrong conclusions. Yet if we account for our natural biases and inclinations and recognize when we (and others) are making assumptions, we will be more likely to evaluate our choices based on the factual data and not merely preconceived notions. Having a healthy suspicion—including misgivings about oneself—is perfectly reasonable and helps us to be as objective as is humanly practicable.

One type of improper inference comes from *circular reasoning*, sometimes referred to as *begging the question*. This argument says, "P is true because Q is true, and Q is true because P is true." If a statement uses one of its premises as a conclusion, it is not a logically coherent argument. The following statement will help clarify the fault of circular reasoning: "Apart from the infallible teaching magisterium of the Roman Catholic Church, evangelicals could never know the extent of the canon of Scripture, nor would they know the Bible is an inspired and infallible revelation from God. Only with an infallible authority can one know definitively that Scripture is the very Word of God. Therefore, the belief in the Bible alone (*sola scriptura*) is a dreadfully absurd proposition because we need the Roman Catholic Church to give us this necessary seal of assurance."

An obvious question comes to mind: how is the authority of the Roman Catholic Church established by the promulgation of the Bible when, according to the argument, the Bible could not be promulgated except by the established church of Rome? This means that if the Catholic church has determined and proclaimed through her infallible rulings just which books are actually part of sacred Scripture and the knowledge of that very canon is incumbent upon the existence of the external church, then how do we explain the presence of the Roman church to produce such a listing? I'm reminded of the perennial question regarding the chicken and the egg: *Which came first?* The circularity of this particular claim is analogous in arguing *the chicken is the consequence of the egg*, while at the same time asserting *the egg is a result of that very same chicken*. Which one is it?

The *slippery slope* argument takes on the following form: if A happens, then by a gradual series of minor steps through B, C, and D, eventually E will happen too. E should never happen; therefore, A should not happen either. Let's look at one culturally relevant canard. "Take away a female's reproductive choice, and the next thing you know it will lead to an unmitigated disaster for all females. If the government can force a woman to continue a pregnancy, what about forcing women to use contraception or undergo sterilization?" This, of course, is sheer nonsense. Pro-life advocates champion the inalienable right to life—a right enshrined in our nation's constitution—and attempt to ensure the weakest of our society are preserved from the earliest stages of development. Talk of compulsory contraception and forced sterilization amount to a red herring and *non sequitor*. There is no compelling reason to believe that upholding the right to life will lead to impositions on any human being.

Double standards can stem from incorrect inferences and can result in confusion. A double standard occurs when one person claims a certain belief or practice for others but not for himself. Some evangelicals, for instance, look for an express warrant or an unambiguous example of infant baptism in the New Testament before they will accept the practice and argue from this basis. But this is a false criterion, which no person could apply consistently. For example, just think about what it would mean for women who wanted to partake of the Lord's Supper. After all, there is no direct command in Scripture to administer the supper to women, nor is there any example to be found in Scripture.[1] What do we do? To answer properly, we must appeal to other passages that admittedly have nothing to do with the Lord's Supper but speak directly to the status of women as equal members in the body of Christ (e.g., Gal 3:28, "There is neither Jew nor Greek . . . male nor female . . . for you are all one in Christ Jesus"). This is a perfectly valid way of doing exegesis and properly interpreting all of God's Word to arrive at a correct conclusion that women have an equal right to the supper.

So when credobaptists charge, "There are no instances of infant baptism in the Bible," they are indeed correct, but the standard is *not* to require explicit examples for every practice. Otherwise, we all find ourselves standing at the crossroads eventually, unable to move forward without clear expressions in Scripture. Demanding anecdotes and illustrations to substantiate every conceivable practice is asking too much; it should not be viewed as the ultimate or only acceptable standard for understanding biblical truths.

The bottom line is that we should not be in the business of using one set of standards to argue in favor of our personal convictions and an entirely different standard

---

1. Some might immediately object by pointing out the "church" and "people" mentioned in 1 Corinthians 11:15–23 undoubtedly includes women in addition to men—even though "women" are never explicitly mentioned. But this only supports my point of applying logical inferences to verses where certain words (e.g., women) may not be present.

when opposing certain practices or beliefs. Consistency is key.

When someone asks more than one question but one is concealed behind the other, that is considered a *loaded question*. "How many more gay people does God have to create before we ask ourselves if he wants them around?"[2] asked one Minnesota legislature rhetorically before the state assembly voted on homosexual unions. By framing the issue in this way, what we have is a representative who is demagoguing the issue of sexual promiscuity by presenting a hidden assumption that God creates people gay; therefore, we must accept them and be tolerant of their lifestyle. The man should not have assumed this. What's more, the gentleman, if he were to remain consistent with his own thinking, could have also asked, "How many more pedophiles (rapists, murderers, etc.) does God have to create before we ask ourselves if he wants them around?" If we applied his logic to other areas, then we would be compelled to accept any and all forms of deviancy.

The fallacy of *equivocation* occurs when someone changes the meaning of the word in the middle of an argument. Sometimes this is done deliberately; other times it is accomplished out of ignorance. An acquaintance of mine once mused: "Of course Augustine believed in the ultimate authority of the Roman magisterium. After all, he wrote, 'I would not believe the gospel unless moved thereto by the authority of the church.'"

Well, yes, quite. But what does Augustine mean by the *church*? Are those who use Augustine's words to justify

---

2. Siedl, "Minn. Legislator: 'How Many More Gay People Does God Have to Create' Before We Ask If He 'Wants Them Around'?"

the authority claims of Rome warranted in pointing to the bishop of Hippo, or is it a flagrant attempt to muddy the waters? Luther believed the latter. When those in communion with Rome circulated Augustine's teaching as "proof" of his concurrence with the authority of the bishop of Rome, the Reformer pointed out Augustine's usage of *church* should not be so severely limited in its scope or application. Luther writes:

> Augustine speaks of the whole Church, and says that throughout the world it with one consent preaches the Gospel and not the Letter of the Manicheans; and this unanimous authority of the Church moves him to consider it the true Gospel. But our tyrants apply this name of the Church to themselves, as if the laymen and the common people were not also Christians. And what they teach they want men to consider as the teaching of the Christian Church, although, they are a minority, and we, who are universal Christendom, should also be consulted about what is to be taught in the name of universal Christendom. See, so cleverly do they quote the words of St. Augustine: what he says of the Church throughout all the world, they would have us understand of the Roman See.[3]

The *either-or* fallacy is when a person insists you must choose between two given options when in reality you have more than two alternatives (briefly touched upon in chapter 3). A false antithesis, as used by skilled propagandists, sets up a position in which either position A is the answer

---

3. Luther, *Works of Martin Luther*, 452.

(which happens to be the position held by the debater) or position B. In this instance, the debater has amply demonstrated the absurdity of B and argues A must be the only option left. However, what the audience might fail to recognize is that C, D, or E might be a better option. Just because B was eliminated does not necessarily make A true since other options are available.

Perhaps you have read the evangelical tract warning us against the detriments of contemporary Christian music—which might more aptly fall under the banner of overzealous criticism. "Music is very important in the spiritual life of a child of God, but rock music, like a raging hurricane, began desecrating the sacred music of the church and hurting Christians in the process. So either you abstain from Christian rock music, in all its present forms, or run the risk of spiritually destroying the church and the children of God."[4]

There is a tendency, when it comes to styles of music in the worship service, to oversimplify the issue, as this tract so ably demonstrated. The amorphous label of "rock music," for starters, is never defined and varies from person to person. Who defines what is and what is not acceptable rock music and upon what basis? Then there are the alleged debilitating effects of certain forms of music and questionable statements given but with no evidence to buttress the charge. The one making the claim has the responsibility of proving his or her case, not making blanket statements and insisting others prove the contrary. Furthermore, no one has the right to play on the emotional fears of others, telling them they are in peril of causing the spiritual demise of an-

4. See Watkins, "Christian Rock: Blessing or Blasphemy?," 7.

other through singing or listening to music that is designed to bring honor and glory to God.

These few illustrations are examples of oversimplification, overreacting, and making assumptions based on scanty or inconclusive evidence—yet many of them are common occurrences, even in our own way of thinking. Guarding our minds and emotions against such strategies will ensure we do not get trapped via misleading information or faulty reasoning.

7

What's the Future Likelihood?

Perhaps you heard about the little boy who brought his father a disturbing report. "Bad news dad," Calvin began as the young lad ambled over to his father, looked down at his notepad, and shook his head in disappointment. "Your polls are way down."

"My polls?" his father asked curiously, peering over his newspaper.

Calvin swept his finger back and forth across the page, as if intently analyzing the latest sample of statistics. "You rate especially low among stuffed animals and six-year-old white males," Calvin responded forlornly. "If you want to stay 'Dad,' I'd suggest you adopt some key planks to your platform. Of those polled, virtually all favor increased allowances and the commencement of driving lessons."

"Well, some interest groups are in for a real surprise," his father mumbled as he returned to his newspaper unaffected by the disturbing "poll numbers."

Adults, much like young Calvin, are just as prone to throw out statistics and numbers when they believe it will help their case. Calvin was hopeful that by giving the impression that the overwhelming majority of "six-year-old white males" surveyed wanted a larger allowance, he would get his way, even though those surveyed represented only one male (a fact conveniently left out of the dialogue). Not to be outdone, grownups do the same thing to get their way; they use data to shape impressions they want to make to advance their own personal agenda. When someone uses the same approach Calvin tried, albeit perhaps a more sophisticated version, we should be leery before making snap decisions or accepting statements without knowing the real numbers behind the number. We don't want to fall prey to any number of statistical fallacies. We'll begin with discussing the *hasty generalization*.

We must be careful to differentiate between *generalizations* and the *hasty generalization*. We all generalize about things; that is, we all make broad comments about people or things based upon our experience and observations. We generalize every day, and this is helpful as we process information and make predictions about someone or something before he or she does it, and sometimes we are right.

A generalization is comprised of samples taken from a *class*, a group of people or things that all have similar attributes or traits that allow them to be classified as such. This can be as broad as male and female or can be more narrowly focused, pertaining to particular belief systems like Covenant theology and Dispensationalism. When you examine one or more people in a particular class, you are taking a sample of that class. A generalization, then, would be to take a sample from any given class and using the information obtained from that sample, to state a conclusion about everyone or everything in that class. When we do not properly sample a class or have insufficient data but still make broad, sweeping statements anyway, we are guilty of making *hasty generalizations*.

"You homeschool your children? Don't you worry about their social development?" I'm sure we've all heard this question or one similar to it—the perennial charge of homeschoolers turning out "socially awkward." Implied here is a prime example of the hasty generalization of assuming all homeschoolers will suffer from some form of social backwardness, yet there is no statistical evidence to support this irrational fear. Like any other skill set, social skills differ from person to person, regardless of formal education. Some people find interacting with others comes naturally while others work hard at feeling comfortable around people. The reality is that we all mature at different stages; some enjoy being surrounded by others while others are naturally more reserved.

Consider the hasty generalization of one of the most prominent journalists during the 1920s: H. L. Mencken. Mencken was noted for his acerbic wit and sarcastic tone, defending secular humanism while attacking Christian fundamentalism, which, according to his view, sought to return America to older and simpler times. Rural evangelicals often received the brunt of his ire. They were a favorite target of his, especially during the infamous 1925 Scopes Monkey Trial in which the teaching of the evolutionary theory was put on trial in Dayton, Tennessee. The following excerpt comes from one of Mencken's voluminous newspaper columns as he covered the trial:

> Let no one mistake [the Scopes Monkey Trial] for comedy, farcical though it may be in all its details. It serves notice on the country that Neanderthal man is organizing in these forlorn backwaters of the land, led by a fanatic, rid of

>   sense and devoid of conscience. Tennessee, challenging him too timorously and too late, now sees its courts converted into camp meetings and its Bill of Rights made a mock of by sworn officers of the law. There are other states that had better look to their arsenals before the Hun is at their gates.[1]

The uneducated "Neanderthals" of whom he speaks are those who rejected the Darwinian evolutionary theory in favor of the biblical account of creation. For Mencken, evangelicals who denied "scientific progress" were deemed crackpots, crazy, and unintelligent. Folks who rejected this revolution in science were people who refused to square the so-called truth of Modernism with the fundamentals of Christian doctrine. By contrast, he regularly spoke of the better-educated and more-sophisticated people who, by his estimation, correctly revoked any and all silly forms of superstition and religion (i.e., those who embraced a Darwinian view of creation).

Mencken's deep-seated convictions ultimately led to tenuous accusations and to false stereotyping, and his hasty generalization against evangelical Christians, referring to them as uneducated *"homo boobiens"*—as rhetorically clever as it might be—was just as improper.

Another faulty form of argument is a *weak analogy*. Analogies are great for making a point or when we want to compare items with each other. When we reason by analogy, we compare two or more items with each other, noticing that these items are, in some manner, the same. Analogies, much like generalizations, can either be strong or weak.

1. Mencken, "The Monkey Trial: A Reporter's Account," 59–61.

When examining analogies, we must examine the items and determine if they are good analogies, and thus strong, or if they fall into the category of being weak analogies. If we scrutinize an argument and determine the differences between the items being compared are major and the similarities are minor, then we can conclude someone is using a weak analogy. Since Islam was catapulted into the national spotlight nearly a decade ago in the wake of the September 11, 2001 terrorist attacks, many have tried to compare Allah of the Quran and the triune God of Scripture, including those within professing Christendom.

According to paragraph 841 of the *Catechism of the Catholic Church*, Muslims, along with Catholics, "adore the one, merciful God." (Vatican II also emphasized that Jews, likewise, worship the one true God.) We read, "The plan of salvation also includes those who acknowledge the Creator, in the first place amongst whom are the Muslims; these profess to hold the faith of Abraham, and together with us they adore the one, merciful God, mankind's judge on the last day."[2] Here we are told that since a group of religious followers embrace the historical figure Abraham as a *rasul*, a prophet, they are included in the plan of salvation. Yes, indeed, it might sound good. But more importantly, is it true?

Admittedly, the Muslim and Christian views of God share some ostensible commonalities. For instance, both believe in one eternal God who created the universe, God as all-powerful, all-knowing, all-present, and so on, but these similarities being to ring hollow in light of the denial of the Son of Abraham. These commonalities are exposed to be bankrupt when we realize those who deny Jesus as

2. *Catechism of the Catholic Church*, 242–43.

the Christ are essentially repudiating the full revelation of who God is as he has revealed himself in human flesh. The entire biblical revelation of the Lord presents us with a picture of a triune God, one being consisting of three persons, namely, God the Father, God the Son, and God the Holy Spirit. The Islamic tradition, by contrast, rejects Christian Trinitarianism (three persons) in favor of Unitarianism (one person), thus denying the incarnation, the revelation of God the Son in the form of Christ.

Jesus revealed he was the only way to eternal life: "I am the way, and the truth, and the life. No one comes to the Father except through me" (John 14:6). Those who refuse to recognize the revealed Son in the person of Christ Jesus and stubbornly decline to follow Christ as Lord and Savior "shall not see life, but the wrath of God remains on him" (John 3:35–36). The God of Abraham, Isaac, and Jacob revealed himself as the triune God of Scripture and in the person of Christ Jesus in the incarnation. Any religious system that rejects this fundamental truth cannot, by definition, be true and cannot make truth claims that stand at open variance with the plain teaching of Scripture. Any religion that makes certain claims that seem comparable to Christianity but cause one to worship a different god is, in the final analysis, still a counterfeit religion.

The Bible states unabashedly that those who do *not* trust in the finished and perfect work of Christ alone will be eternally separated from the Lord. To deny Jesus, his incarnation, his sinless life, his atoning death, his resurrection, and his ascension into heaven is to be repudiated. And Islam does all these things. Not only does Islam *not* look to the work of Christ for the atonement of sins, but it also does not teach a message of divine grace. In fact, one cannot truly worship the true and living God of Scripture while denying his Son. If the Bible's message is true, then Islam cannot be true; Muslims cannot worship the same God as Christians, despite the strongest assertions of the Roman Catholic Church and the popes. If we are to take the Bible seriously and suppose the Christian faith is true, then we must conclude any religion that denies these essential truths is a false religion.

The *proof by lack of evidence* is a fallacy that claims something is true solely because nobody has proffered sufficient evidence to the contrary. For example, someone might say, "There are no passages in Scripture that directly state that Mary was *not* conceived without original sin or that she was *not* immaculately conceived. Therefore, the church's teaching regarding the sinlessness of Mary is perfectly compatible with Scripture."

If Paul's words that "all have sinned and fall short of the glory of God" (Rom 3:23) are not clear enough, and we are to believe that the Bible must make positive declarations as to the sin of every specific character mentioned therein, then we must conclude that more individuals than simply Mary lived a perfect and sinless life. The Bible never mentions a sin by Daniel or Job, but we still assume they sinned like the rest of us. Going further, the Bible doesn't say Priscilla, Lydia, Clement, Tertius, Lucius, Jason, Asyncritus, Patrobas, or Gaius (among a host of others) were conceived without original sin or that they were not immaculately conceived, yet we do not assume that any one of them was. A paucity of evidence does not support or further an argument.

Post hoc ergo propter hoc is a Latin phrase meaning, "after this, therefore, this." Essentially what this fallacy states is that since X happened before Y, then X must have caused Y. Consider this example: "Like many of us, I was stuck in a job where I would arrive at work, sit behind a computer screen for eight hours a day, and then go home. Travel is my passion, as it has always been, and I longed to venture all over the world. I didn't know what to do until my friend suggested I ask St. Therese to help and to intercede on my behalf. I took her advice and prayed to St. Therese. During my prayer, I recalled her words that she will 'let fall from heaven a shower of roses,' so I asked her to bring me a rose to let me know if I should continue working for this particular airline. I finished my prayer, and about an hour went by before the secretary in our department came into my office and handed me a rose. I was speechless; I couldn't believe what had just happened. Needless to say, I listened to the answered prayer of St. Therese, stayed at my job, and received a promotion and the opportunity to travel abroad. There is no doubt in my mind that saints hear our prayers and intercede on our behalf, just as St. Therese answered my prayer."

Of course, it doesn't necessarily follow that because this lady prayed to a deceased saint and subsequently received a rose and a promotion that her prayer was the immediate cause of the temporal blessing. It is also very possible that she received a rose from one of the many rose bushes directly outside her office, that someone was displaying a random act of kindness, or that her promotion was the result of her tireless work ethic, among a number of other more plausible scenarios. If, however, X happened before

Y, we should not immediately conclude that X caused Y; we simply need to find out more details before reaching dogmatic conclusions.

A variation of this fallacy is the closely related *post hoc ergo propter hoc in statistics*. The following article appeared online in the *Rochester Independent Examiner*.

> Once again, science trumps superstition and ignorance.
>
> At one time, people believed that the Earth was the center of the universe. Then science proved them wrong. People believed the Earth was flat, until science proved them wrong. People also used to believe that bleeding a person was good for their health, but science proved that wrong too.
>
> Science has proved conclusively that human activity is causing global warming, but some conservatives still don't believe it. Maybe they also still believe that the Earth is the center of the universe, that the world is flat, and that bleeding a person is good for their health.[3]

Always be leery of the "obvious" conclusion—in this instance, that science conclusively determines human activity is the immediate cause of climate change and increasing temperature. The article never mentioned which scientific data proves this to be factual; we are merely told to believe it. Why? Because smart people with copious statistics and data say so.

---

3. Mangan, "Global Warming Timetable Proves Skeptics Wrong," 1–12.

Of course, there is no proof that certain human activities have the capacity to change the environment of a universe fashioned by the very breath of Almighty God. The reality might show that the Earth goes through natural cycles of temperature change that are independent of human activity and perhaps a bit more cynically, that those sounding the alarm against "excessive carbon emissions" might have ulterior motives for concocting yet another manmade calamity. For those old enough to remember back to the 1970s, much ado was made in the media over *global cooling*, yet it turned out to be broadcasting hype.

All of these statistical fallacies might give the initial impression of veracity or truthfulness, but before anyone accepts statements as fact based solely on numbers or data, he or she should understand the information in context to get a proper understanding of the information being presented. If someone is not willing to do this, then it might very well expose a fraudulent attempt at manipulating the data and the audience.

8

If Only It Were That Simple

"Whiskey is for drinking—water is for fighting over," quipped Mark Twain, America's most famous literary icon. Much like his legendary quote, discourse in our daily lives tends to be oversimplified. We see this approach as a quick, easy solution to complex dilemmas when we don't really want to be bothered with thinking through the difficulty of our opponent's position or if we simply don't want to be concerned with the possible ramifications of an issue. In previous chapters, we have touched upon some forms of oversimplification, but here are additional types of oversimplification that are ever-present in daily discourse.

The first form of oversimplification is the fallacy of *accident,* which occurs when someone applies a general rule to a specific situation in which it is not intended. This fallacy suggests all situations are black-and-white circumstances in which there are no exceptions to the rule. For instance, some religious pacifist groups cite the commandment, "Thou shalt not kill" as condemnation against military warfare. Similarly, people who are opposed to capital punishment cite this commandment to accuse Christians of being hypocritical if they support the death penalty imposed upon a person by judicial process as a punishment for an egregious crime against humanity. In both instances, though, the objector is committing the fallacy of accident.

Pacifists myopically seem to focus on the letter of the law while ignoring the spirit of the law. However, a thorough reading of the Bible leads one to the opposite conclusion of pacifism. Beginning in the Old Testament, we see multiple occasions in which the Lord commanded Israel to conquer her enemies, punishing pagan nations for their wickedness—and conversely permitting Israel to be subjugated at times for their rejection of him (Num 21:3, 35; 25:4; 31:7–9; 31:17–18; Deut 2:33–34; 7:1–2; 20:16; Jos 8:22–25; 10:10–43; cf. 1 Chr 9:1). Other examples of divinely sanctioned capital punishment include the destruction of Sodom and Gomorrah (Gen 18:20—19:26), death of the firstborn in all of Egypt (Exod 11:4–5), and punishment for the seditious Israelites involved in Korah's rebellion (Num 16:1–40). The New Testament likewise advocates for civil authorities using the sword. In Revelation 19, we see perhaps the strongest example of military prowess in the portrait of Jesus, the warrior King. Jesus returns to wage war against his enemies, treading out the winepress and annihilating his unrighteous enemies by spilling their blood over his robe—scarcely a picture of a subdued pacifist.

The *complex question* occurs when someone raises an issue or presents a conundrum that has several consequences but whose ramifications are ignored or not recognized. "Is Islam a religion of peace—yes or no?" The crude response desired tries to boil this down to a yes or no answer, but there are too many parts to the proposal to answer in such a simplistic way. The complex question oftentimes amalgamates more than one question into a basic statement or question: "Are you still rebelling against God's divinely instituted authority?" Depending on who is asking the question and in what context, this form can entail at least two separate questions: "Are you rebelling now? Have you ever rebelled against the divinely instituted authority?" Then we could add: "Is there a divinely instituted authority we are to obey? What is that organization, and what should it look like?"

Failure to recognize the complex question leads to all sorts of confusion and misunderstanding of an issue. Then there are times when the complex question takes on the form of a statement. Note the following thesis for an evangelical debate: "Gay marriage is consistent with New Testament Christianity." Masquerading as a truth claim, this proposition really has multiple resolutions:

(1) that marriage has *not* been exclusively defined by God as the union between one man and one woman;

(2) that same-sex couples are permissible according to the New Testament. If we understand that at times a singular question includes other assumptions or questions that must first be addressed, we will be better equipped to provide a meaningful response.

The *excluded middle* is known as the black-and-white fallacy, and it is the mistake of assuming something is *either* one thing *or* another. "Either you affirm a literal six-day creation narrative in Genesis or you cast doubt, albeit perhaps unwittingly, on the inspiration of divine Scripture." Actually, some may support the plenary inspiration and inerrancy of holy writ and still question the historic understanding of Genesis 1 and 2. Evangelicals committed to preserving the highest integrity for God's Word differ widely on the medium God used in the creation process.

"Either we vote with conservative Republican candidates or we vote to kill the unborn." While it is difficult for any consistent Christian to knowingly vote for someone who overtly supports the "right" of another human to destroy life under the guise of reproductive rights, there are other options that should not be ignored. There are degrees of freedom and conscience, including, but not limited to, the freedom *not* to vote or the liberty to vote for a candidate who is affiliated with a party other than one from the two established political platforms. This fallacy, inherent in most slogans or bumper stickers, commits errors of the most extreme type, generally reducing a situation to only one of two options.

Pigeonholing is similar to the excluded middle. This misguided approach occurs when we put forth our own personal viewpoint on any particular issue. Sometimes in an effort to neatly categorize a complex issue, we oversimplify a subject and force it into a category into which it does not belong. A colleague remarks, "If you're not a Calvinist, then you must clearly be an Arminian." Though it might be easier to classify believers as either-or, evangelicals might not fully embrace classical Arminianism. They might embrace one, two, or more traditional points of Arminian theology, but this does not automatically make someone an Arminian. Instead, perhaps a more suitable distinction in the age-old debate would be to classify Christians as *monergists* (God alone saves) and *synergists* (God, along with man's cooperative effort, brings about salvation).

One of the chief results of oversimplifying is *jumping to conclusions*. I remember on multiple occasions listening to an evangelical accuse Roman Catholics of worshipping Mary, only to be rebuked by those in communion with Rome for misunderstanding Roman theology—and correctly so, I might add. These evangelicals made spectacular claims they were unable to support because they were not aware that Rome teaches the veneration of Mary and the saints differs, according to their understanding, from the worship of God. According to official teaching, Roman Catholics assert worship given to God is called *latria*, veneration offered to the saints is called *dulia*, and the honor paid to Mary is referred to as *hyperdulia*. According to Rome, the highest form of worship is reserved solely for God, whereas a lesser form is offered to Mary and the saints. In the Roman Catholic perspective, they are not worshipping Mary as they would God. Beginning a conversation with accusatory tones of idolatry does not permit an environment for meaningful dialogue and interaction.

Absolutes are a common form of oversimplification. People love to make dogmatic, unyielding, and firm statements of fact: *all*, *everyone*, *without exception*, *always*, *never*, *no one*, and so on and so forth. One famous preacher created some stir in certain evangelical circles after he preached from 2 Corinthians 5:14, a verse that reads, "For the love of Christ compels us; because we judge thus, that if One died for all, then all died" (NKJV). The preacher proceeded to explain that this passage unequivocally teaches Christ died for all humanity without exception. He recalled the words of his hermeneutics professor, who taught, "The word 'all' means all, and that's all the word 'all' means."[1]

Of course, we know that *all* does not always mean *all humanity without exception*, because basic rules of exegesis remind us that the immediate context defines how the word is to be understood—and sometimes this means words are limited in scope, including the word *all*. For instance, when Luke records that a decree went out that "*all* the *world* should be registered" (2:1), *all* and *world* were limited to represent denizens under the authority of the Roman Empire and not native inhabitants of Asia, Africa, or the continents of North and South America.[2] Similarly, *all* in the Corinthians passage does not necessarily have such an overreaching usage. The verse does not automatically insist we interpret this verse to mean "Christ's death redeemed all of humanity without exception—bar none." In fact, the two occasions of *all* in the Corinthians verse should be understood to mean "all for whom Christ died—the elect—also died." After all, since those for whom Christ died also died in Christ, the very same must necessarily infallibly live "in Christ." Since this is the case, the possibility that someone

---

1. For an expanded explanation of this position and a rebuttal of Reformed theology, see Norman Geisler's sermon entitled "Why I am Not A 5-Point Calvinist" at http://media.calvaryftl.org/player/?fn=G5146.

2. The Greek term *kosmos* (translated into English as *world*) has at least seven clearly defined different meanings in the New Testament. For instance, note the differentiation between the *world of unbelievers* and the *world of believers*. "If the world hates you, know that it has hated me before it hated you" (John 15:18); and, "For the bread of God is he who comes down from heaven and gives life to the world" (John 6:33). For further discussion, see Pink, *The Sovereignty of God*, 289–92.

who died in Christ is not raised from the dead is as unlikely as thinking Christ might have remained in the grave.

Those for whom Christ died, and subsequently those who died in Christ, are necessarily renewed individuals who will be spared from eternal death. It is these people who are being mentioned by the apostle Paul in Corinthians as the "all" of the verse since we know that *all* humanity does not see "everlasting life"—only those who believe in Christ. Interpreting the *all* consistently in 2 Corinthians to mean *all humanity without exception* leads us to the untenable position of universalism, a position few evangelical believers would sanction.

Then there are the provocative statements assaulting Christian beliefs from a wide spectrum of sources including newspapers and Internet articles to books and public lectures that grossly oversimplify issues and only serve to patronize the opposition. I'm reminded of an article appearing in the *UK Times* written by Oxford Professor and outspoken atheist, Richard Dawkins, entitled: "Creationism: God's gift to the ignorant." That, needless to say, pretty much says it all, doesn't it? So much for fair and balanced. In another instance, Sam Harris, another celebrated atheist, remarked, "All religious adherents must be exposed to the illegitimacy of their core beliefs." Elsewhere, "The three Abrahamic religions in the world today can be traced back to the Holy Land, the least enlightened place in the world, a microcosm of the threat to rational values and civilization posed by religion, whose irrational roots are nourishing intolerance and murder." Again, I point out, we have an oversimplification of the matter.

If Only It Were That Simple 111

In each instance, you might have noted, no positive evidence was presented as to why evolution is superior to Intelligent Design or why atheism provides a better moral framework than does, say, Christianity. Be wary of statements that use absolute terms, because things are not always as simplistic as some people would have us believe.

The *false mean* is also referred to as the *fallacy of compromise*. Understandably, there are times when cooperation between two opposing forces is necessary for the common good, but this is not always the case. In fact, sometimes compromise can lead to precarious ends. "I'll promise you this, preacher," remarks the congregant. "I'll continue to live my hedonistic lifestyle for just a few more years, and then I'll seek Jesus and repent of my sins." As unlikely as this scenario is, the thrust of it crystallizes how some people play Russian roulette with their lives, convincing themselves they will change for the better at some future date. In this instance, compromise is not desirable—at least for the one who wants eternal life.

One particular ecclesiastical conference wants to ordain homosexual ministers; another feels doing so is against the clear teaching of Scripture. A compromise that the established churches ordain openly gay ministers in a more liberal region of the country, or some similar cooperative effort, is unacceptable if, indeed, sanctioning the homosexual lifestyle is intrinsically wrong.

Sometimes politicians, in an attempt to appease social conservatives and liberal activists, openly oppose gay marriage, yet at the same time, in a display of compromise and mollification, permit homosexual unions. This politically calculated compromise sprouting up throughout the

United States is just as intolerable as the original proposal. If something is fundamentally wrong, it is wrong, however much justification some try to give it.

The *fallacy of the fall,* as one might expect, references the fall of humanity in the Garden of Eden, pointing out that we are all imperfect and flawed human beings. In reality, it is more of a rationalization for bad behavior. "Don't be too hard on the boy," the mother pleads. "No one is perfect." Chances are that the one who makes this statement might not want to be bothered with taking some corrective action or imposing discipline.

Another oversimplification is the *fallacy of determination.* Imagine this scene: There is a paraplegic in your church, someone who has suffered the consequences of a debilitating illness since he was three, and people are telling him he doesn't have enough faith; otherwise God would have healed him. "The devil is a liar," the prosperity teacher storms. "Healing is the children's bread. It's promised to us in the Word of God; God can do things medicine cannot do." People remind him constantly that the Lord would heal him if his faith were stronger. Can you imagine what happened in this real-life situation? As you might expect, his lingering condition caused him to question the very existence of God.[3]

This fallacy suggests that anything is possible and probable. Granted, we understand that, in fact, everything with the Lord is a possibility. But this error states that the primary reason you do not have something is because you do not want it enough. If something hasn't happened to you yet, it's because you don't want it strongly enough to make

3. See Hanegraaff, "If You Are not Healed, Do You Lack Faith?"

it happen. In other words, you haven't been determined enough to bring about your own outcome (i.e., you have not tried hard enough or possessed enough faith to be healthy, wealthy, or wise). Hard work and determination do play an integral role in bringing about temporal blessings, but one should not automatically presume upon the goodness and grace of God simply because one puts forth enough faith or effort. Sometimes the Lord's answer is the same one he delivered to the apostle: "My grace is sufficient for you" (2 Cor. 12:9).

The *fallacy of idealism* is another form of oversimplification. "You don't spank children today; in the old days, maybe we got spanked, but there was a different quarrel. You don't spank children. You understand?"[4] These were the words uttered to a Texas mother by a Corpus Christi judge who sentenced the mother of three children to five years' probation, a $50 fine, and parenting classes for spanking her two-year-old daughter on the rear end. Instead of recognizing the value of corporal punishment, the judge proposed that the mother find alternative ways of correcting her toddler, such as rationalizing with the toddler or meeting with her halfway, as if to help her see the folly of her ways. Then, in the long run, she would come to see how her behavior, if not checked early, would lead to destructive practices later in life.

This, in essence, was the tripe the judge was advocating. Such glibness is often used by those who have limited experience or deficient wisdom to approach a complex problem in a constructive and experienced way. Often those who think they know best offer little more than sub-biblical

4. Anonymous, "Mom gets probation for spanking child," 10–11.

and imprudent advice, leading observers to question their judgment. The ancient Proverb rings true: "Whoever spares the rod hates his son, but he who loves him is diligent to discipline him" (13:24). Imposing "chat sessions" upon parents in lieu of the rod when needed is, as the Bible teaches, pure folly.

The *fallacy of tacit agreement* states that since no one has overtly protested, everyone must, therefore, agree. In truth, there are plenty of reasons people do not always voice opposition. The fact that the speaker has not heard dissent does not mean there is no dissent or disagreement among the listeners. Some may be shy; others might be afraid of looking foolish; or still others might be intimidated by the audience reaction or of the potential consequences from the speaker—for example, the one who has the authority to assign a failing grade to an incorrigible student or the person who can take away one's livelihood for not conforming (just as we have witnessed repeatedly in the scientific community with those who embrace Intelligent Design).[5]

---

5. See Limbaugh, *Persecution*. See also Frankowski, *Expelled*.

9

Adieu To You and You and You

In his bestselling book, *Seven Habits of Highly Effective People*, author Stephen Covey recounts a poignant encounter with a stranger on a subway in New York that forced him into a "mini-paradigm shift." He writes about the calm, peaceful scene of the train ride that Sunday morning when all of the sudden a man and his children entered the subway car and instantly altered the tranquil atmosphere. The children were rambunctious, running wild, grabbing newspapers from people, and causing a ubiquitous feeling of irritation among the passengers.

Covey was perturbed with the high-spirited children but was even more vexed with the father's seeming insensitivity to and irresponsibility for the conduct of his children. Finally, and with a sense of righteous indignation, he turned to the man and pointed out how his children were disturbing a lot of people and asked if he ever considered controlling them a bit more.

The father looked up at Covey for the first time during the train ride and as if returning to consciousness from a deep slumber, answered in a hushed tone, "Oh, you're right. I guess I should do something about it. We just came from the hospital where their mother died about an hour ago. I don't know what to think, and I guess they don't know how to handle it either."[1]

You can only imagine the horror and sinking feeling Covey experienced at that very moment. Instantly, Covey remembers, his entire way of thinking changed; his paradigm shifted. He suddenly *saw* things differently based on that encounter, and because he *saw* things differently, he *thought* differently, he *felt* differently, and he *behaved* differently. For the first time, he was able to think differently because of a transformative experience in his life and was able to reevaluate his priorities in light of a new situation.

Like Covey's experience, understanding some of these logical fallacies might just lead us into our own mini-paradigm shift, a renewed way of looking at the evidence presented before us prior to reaching certain conclusions. If we remember some of the most important principles encapsulated in the preceding chapters, we will be well on our way to making informed, objective conclusions in our journey into discovering *true truth*. So to summarize what has been illustrated throughout this book, these are some of those notable principles in a nutshell:

1. Be wary of those who speak in absolutes; things are never that simple.

2. Watch out for unsubstantiated generalizations.

3. Emotional language can be highly persuasive and lead us into bad decisions; be on guard.

---

1. Covey, *The Seven Habits of Highly Effective People*, 31.

4. Remember not to confuse empirical evidence with feelings, thoughts, biases, attitude, speculation, or personal opinion.
5. Be sure that both sides understand the main issue under consideration, and do not deviate from the topic needlessly.
6. Ensure the arguments for or against a position are directly related to the topic of discussion.
7. Might does not make right; just because someone says so does not automatically make it so.
8. Always ensure the conclusion logically follows from the stated premises.
9. Do not make unwarranted inferences from your own biases or prejudices; do not make assumptions, because oftentimes they are wrong.
10. Always remember to give an answer with gentleness and respect. If your passions are overshadowing your message, then your approach must be reevaluated.
11. Make use of all the data, and do not cherry-pick information that fits your side.
12. Do not engage in debate for the mere sake of debate. Dialoguing should be constructive and ultimately for the glory of God, to bring others the good news about Christ Jesus.
13. Think rationally about all issues in life, not merely theological concerns. God cares about every facet of our lives, and we should seek to honor God in word, in thought, and in deed.

14. Whenever you encounter an argument, remember to examine it critically in light of the information you have learned, asking yourself if the premises are true, if the conclusion naturally follows from the given premises, and if all the evidence has been presented for consideration.

15. Finally, remember that God's Word is *the* final authority in all matters of faith and practice. Because the Bible is literally "breathed-out by God" (2 Tim 3:16), we can have no higher authority than what the apostle told us is God's very Word. "Sanctify them in the truth; your word is truth" (John 17:17).

The sobering reality is that reason is under assault every day in our society, and we must be ready and equipped to fight the good fight without fear of reprisal. We need clear-minded Christians providing reasoned defenses of the faith, believers who are able to give an answer to every person. We need Christians who give logical, intelligent, rational explanations for the truth of Christianity to people who claim the truth of the triune God of Scripture to be folly. There are far too many who are ready and willing to undermine the teachings of Jesus. The time to express a reasonable faith is now, before it is too late. We must all be in a position of *Reclaiming Reason*.

Summary of Fallacies and Rubbish

CHAPTER TWO: THE LANGUAGE OF EMOTIONS (EMOTIONAL LANGUAGE)

> Appeal to Pity
> Plea for Special Treatment
> Appeal to Guilt
> Appeal to Fear
> Appeal to Flattery
> Appeal to Hope
> Appeal to Sincerity
> Appeal to Friendship/Love/Trust
> Appeal to Pride/Loyalty
> Appeal to the Bandwagon
> Appeal to Status
> Appeal to Tradition

CHAPTER THREE: THE QUEST TO SHAPE PUBLIC OPINION (PROPAGANDA)

> Repetition
> Confidence
> Oversimplification
> Name-Calling or Labels
> Poisoning the Well
> Stereotyping

The Glittering Generality
Slogans
Transfer
Testimonial
Statistics without Context
Arrant Distortion or Stacking the Deck

Chapter Four: What Does That Have To Do With the Price of Tea In China? (Irrelevance)

Ad Hominem
Guilt by Association
Tu Quoque
Counter-Question
Irrelevant Reason
Non Sequitur
Irrelevant Details
Appeal to Force
Appeal to Ignorance
Vague Appeal to Authority
A Priori
Sacred Cow
Jargon

Chapter Five: A Straw Man Never Fights Back (Diversion)

Red Herring
Straw Man

Chapter Six: Why Assumptions and Incorrect Inferences Get Us Into Trouble (Assumptions and Incorrect Inferences)

> Circular Reasoning/Begging the Question
> Slippery Slope
> Double Standards
> Loaded Question
> Equivocation
> Either-Or

Chapter Seven: What's the Future Likelihood? (Statistical Fallacies)

> Hasty Generalization
> Weak Analogy
> Proof by Lack of Evidence
> *Post Hoc Ergo Propter Hoc*
> *Post Hoc Ergo Propter Hoc* in Statistics

Chapter Eight: If Only It Were That Simple (Oversimplification)

> Accident
> The Complex Question
> The Excluded Middle
> Pigeonholing
> Jumping to Conclusions
> Absolutes
> The False Mean/Fallacy of Compromise
> Fallacy of the Fall
> Fallacy of Determination
> Fallacy of Idealism
> Fallacy of Tacit Agreement

Bibliography

*The following titles influenced the preparation
of this book in one way or another.*

Anonymous. "Christian preacher on hooligan charge after saying he believes that homosexuality is a sin." *Mail Online*. Online: http://www.dailymail.co.uk/news/article-1270364/Christian-preacher-hooligan-charge-saying-believes-homosexuality-sin.html.

Anonymous. "Justice Roy Moore's Lawless Battle." *New York Times*. Online: http://www.nytimes.com/2003/08/13/opinion/justice-roy-moore-s-lawless-battle.html.

Anonymous. "Mom gets probation for spanking child," *Eyewitness News*. Online: http://abclocal.go.com/ktrk/story?section=news/state&id=8195860.

Bainton, Roland. *Hunted Heretic: The Life and Death of Michael Servetus 1511 – 1553*. Boston: Beacon Press, 1960.

Baker, Samm Sinclair. *The Permissible Lie: The Inside Truth About Advertising*. Boston: Beacon Press, 1968.

Beardsley, Monroe C. *Practical Logic*. Englewood Cliffs: Prentice-Hall, 1950.

Bell, Rob. *Love Wins: A Book About Heaven, Hell, and the Fate of Every Person Who Ever Lived*. New York: HarperOne, 2011.

Bettenson, Henry and Chris Maunder, eds., *Documents of the Christian Church*. 3d ed. New York: Oxford University Press, 1999.

Bluedorn, Nathaniel and Hans Bluedorn. *The Fallacy Detective: Thirty-Six Lessons on How to Recognize Bad Reasoning*, 2d ed. Muscatine: Christian Logic, 2003.

Brauch, Manfred T. *Abusing Scripture: The Consequences of Misreading the Bible*. Downers Grove: InterVarsity Press, 2009.

Brown, J. A. C. *Techniques of Persuasion: From Propaganda to Brainwashing*. New York: Pelican Books, 1977.

Brown, Michael L. *A Queer Thing Happened to America: And what a long, strange trip it's been*. Concord: Equal Time Books, 2011.

Bulwer-Lytton, Edward. *Richelieu: or, The Conspiracy*. New York: Dodd, Mead, and Co., 1896.

Carson, D. A. *The King James Version Debate: A Plea for Realism*. Grand Rapids: Baker, 1978.

Catechism of the Catholic Church, 2d ed. New York: NY, Doubleday, 2003.

Chan, Francis. *Erasing Hell: What God said about eternity, and the things we made up*. Colorado Springs: David C. Cook, 2011.

Comes, Lisa. "Can You Perceive It?" Joel Osteen Ministries. Online: http://www.joelosteen.com/HopeForToday/ThoughtsOn/Finances/CanYouPerceiveit/Pages/CanYouPerceiveit.aspx.

Coulter, Ann. *Demonic: How the Liberal Mob is Endangering America*. New York: Crown Forum, 2011.

Covey, Steven R. *The 7 Habits of Highly Effective People: Powerful Lessons in Personal Change*. New York: Free Press, 2004.

Frankowski, Nathan. *Expelled: No Intelligence Allowed*. DVD. British Columbia: Premise Media, 2008.

Foreman, Martin. "Problems With God: Disease and Disaster." Online: http://www.godwouldbeanatheist.com/2problem/207disease.htm.

Fudge, Edward. *The Fire That Consumes: A Biblical and Historical Study of the Doctrine of Final Punishment*, 3d ed. Eugene: Wipf & Stock, 2011.

Geisler, Norman. *Chosen But Free: A Balanced View of Divine Election*, 2d ed. Minneapolis: Bethany House, 2001.

———. "God Knows All Things." In *Predestination and Free Will: Four Views of Divine Sovereignty and Human Freedom*. Downers Grove, IL: InterVarsity, 1986.

Gerstner, John. *Repent or Perish: A Biblical Response to the Conservative Attack on Hell*. Ligonier, PA: Soli Deo Gloria, 1990.

Gonzalez, Justo L. *The Story of Christianity: The Reformation to the Present Day.* New York: Harper, 1985.

Gula, Robert J. *Nonsense: A Handbook of Logical Fallacies.* Mount Jackson: Axios Press, 2002.

Hagin, Kenneth. *Seven Things You Should Know About Divine Healing.* Broken Arrow: Faith Library, 1979.

Hanegraaff, Hank. "If You Are not Healed, Do You Lack Faith?" Christian Research Institute. Online: http://www.equip.org/hank_speaks_outs/if-you-are-not-healed-do-you-lack-faith-.

Hendryx, John. "Arminian Suicidal Tendencies: How to Answer the Arminian Charge that Calvinism is Fatalistic." Online: http://www.monergism.com/thethreshold/articles/onsite/lavender.html.

Hitchens, Christopher, *God is Not Great: How Religion Poisons Everything.* New York: Twelve, 2007.

Jefferson, Thomas. "To Thomas Paine Philadelphia, June 19, 1792." Online: http://odur.let.rug.nl/~usa/P/tj3/writings/brf/jefl99.htm.

Lepicier, Cardinal Alexis Henry Marie. *Indulgences, Their Origin, Nature, and Development.* London: Kegan Paul, Trench, Trubner, & Co., 1895.

Limbaugh, David. *Persecution: How Liberals Are Waging War Against Christianity.* New York: Perennial, 2004.

Luther, Martin. *Works of Martin Luther*, Vol. 2, eds. Jacobs, Henry Eyster and Adolph Spaeth. Philadelphia: A. J. Holman and The Castle Press, 1915.

Mangan, Thomas. "Global Warming Timetable Proves Skeptics Wrong," *Rochester Independent Examiner.* Online: http://www.examiner.com/independent-in-rochester/global-warming-timetable-proves-skeptics-wrong.

Mencken, H. L. "The Monkey Trial: A Reporter's Account." Online: http://law2.umkc.edu/faculty/projects/ftrials/scopes/menk.htm.

Moo, Douglas J. *The Epistle to the Romans,* The New International Commentary on the New Testament series. Grand Rapids: Eerdmans, 1996.

Murrell, Adam. *So You Want to Become a Roman Catholic?: Ten Letters You Must Read Before Leaving the Evangelical Faith.* Amityville: Great Christian Books, Forthcoming.

Nance, James and Douglas Wilson. *Introductory Logic: For Christian and Home Schools.* Moscow: Canon Press, 2006.

Olson, Theodore Olson. "The Conservative Case for Gay Marriage." *Newsweek Magazine*. Online: http://www.thedailybeast.com/newsweek/2010/01/08/the-conservative-case-for-gay-marriage.html.

Pink, A. W. *The Sovereignty of God*. Alachua: Bridge-Logos, 2008.

Pusey, Edward. *Is Healthful Reunion Impossible? A Second Letter To The Very Reverend J. H. Newman*. London: James Parker & Co., 1870.

Rowell, J. B. *Papal Infallibility: It's Complete Collapse Before a Factual Investigation*. Grand Rapids: Kregel Publications, 1963.

Saad, Lydia. "Americans Believe Religion is Losing Clout: Percentage saying influence of religion is slipping at 14-year high." Online: http://www.gallup.com/poll/113533/americans-believe-religion-losing-clout.aspx.

Seidl, Jonathon. "Minn. Legislator: 'How Many More Gay People Does God Have to Create' Before We Ask If He 'Wants Them Around'?," *The Blaze*. Online: http://www.theblaze.com/stories/minn-legislator-%E2%80%98how-many-more-gay-people-does-god-have-create-before-we-ask-does-god-actually-want-them-around%E2%80%99.

Shakespeare, William. *Hamlet: Prince of Denmark*. No city: Plain Label Books, 2009.

Sproul, R. C. *Essential Truths of the Christian Faith: 100 Key Teachings in Plain Language*. Carol Stream: Tyndale, 1998.

Storms, Sam. "How Can God Be Loving?" Enjoying God Ministries. Online: http://www.enjoyinggodministries.com/article/how-can-god-be-loving.

Watkins, Terry. "Christian Rock: Blessing or Blasphemy?". Christians for Truth. Online: http://www.cft.org.za/articles/rockmusiccrock.htm.

Watts, Isaac. *Logic or the Right Use of Reason in the Inquiry After Truth*. Whitefish: Kessinger Publishing, 2004.

White, James and Bart Ehrman. "Can the New Testament be Inspired in Light of Textual Variation?" Online: http://mp3.aomin.org/805Transcript.pdf.

James, White. "Sola Scriptura in Dialogue." Alpha and Omega Ministries. Online: http://vintage.aomin.org/SS.html.

———. *The King James Only Controversy: Can You Trust the Modern Translations?*, exp. ed. Minneapolis: Bethany House, 2009.

———. *The Potter's Freedom: A Defense of the Reformation and a Rebuttal of Norman's Geisler's Chosen But Free*, rev ed. Amityville: Calvary Press, 2009.

Wilson, Douglas. *To a Thousand Generations: Infant Baptism—Covenant Mercy for the People of God*. Moscow: Canon Press, 1996.

Wilson, Douglas and Christopher Hitchens. *Is Christianity Good for the World?* Moscow: Canon Press, 2008.

www.ingramcontent.com/pod-product-compliance
Lightning Source LLC
Chambersburg PA
CBHW050830160426
43192CB00010B/1975